Rye's Battle of the Century

Saving the New Hampshire Seacoast from Olympic Oil

We abuse land because we regard it as a commodity belonging to us. When we see land as a community to which we belong, we may begin to use it with love and respect.

Aldo Leopold

Rye's Battle of the Century

Saving the New Hampshire Seacoast from Olympic Oil

Lisa Moll

Rye Historical Society
Rye, New Hampshire 03870
2016

ISBN: 978-0-692-69208-0

Library of Congress Control Number: 2016941874

Published by
Rye Historical Society
PO Box 583
10 Olde Parish Road
Rye, New Hampshire 03870

www.ryenhhistoricalsociety.org

Book design: Grace Peirce

More information on the Onassis Olympic Oil battle may be found
on the Web at www.ryenhhistoricalsociety.org

Printed in the United States of America
First Printing, 2016

Contents

Foreword

THIS BOOK WAS INSPIRED by town of Rye resident Lisa Moll's University of New Hampshire research paper entitled "Rye's Ode to Olympic Oil," which demonstrated the crucial role Rye played in stopping Olympic Oil's effort to build the largest oil refinery in the world on Great Bay in Durham, New Hampshire. Rye blocked the Olympic effort to secure a marine terminal on the New Hampshire Isles of Shoals to receive crude oil and the pipelines needed to transport oil for refining inland. Lisa's paper, part of which was researched at the Rye Town Museum, also provides a full overview of the role of Durham and other seacoast towns in the defeat of the refinery.

The Rye Historical Society (RHS) is grateful to all the people who fought tirelessly to save the coastline of New Hampshire from exploitation. Particular thanks is given to the late Guy Chichester of Rye who fought the proposal tirelessly and donated his collection of material to the Town museum, and to Jessie Herlihy, founder of the Rye Historical Society in 1976, who held anti-refinery meetings in her home in 1973-74.

On the fortieth anniversary of the oil refinery defeat in 2014, Dudley Dudley, who helped spearhead the defeat of the proposal in Durham and in the New Hampshire legislature, gave a talk in Rye. At that time, Peter Horne, one of the key Rye activists against the refinery, spoke about his role in the defeat. Peter's reflections are included in this book, along with other Rye activists.

RHS would also like to thank the following people who contributed financially to the publication of this book: the Peter and Holly Horne family, Marion and Tom Barron, Cynthia Muse, Betsy and Bill Bischoff, Mike and Sally Flanigan, Pete and Nancy Winthrop, the Rye Historical Society, and Alex Herlihy. To author Lisa Moll, RHS owes a great debt for her tireless work in bringing all the parts of this book together. RHS is also grateful to Grace Peirce for her professional services in bringing this book to print.

The Rye Historical Society is glad to play a part in telling this crucial story of town and regional history. Its lessons for today in fighting inappropriate, brute force technology are readily apparent.

Alex Herlihy
Rye Historical Society/Town Museum
2016
www.ryenhhistoricalsociety.org

Preface

IN THE WINTER OF 2015, I began researching New Hampshire's environmental history as part of my graduate studies at the University of New Hampshire. My initial line of inquiry raised more questions than answers. In time, disparate pieces of information, people, and events, clarified into one compelling piece of New Hampshire's history.

This essay has been crafted out of newspaper clippings, photographs, handwritten notes, and author interviews. I have focused on the town of Rye to expand our understanding of this important part of New Hampshire's history. People and events that are less well known have been brought into the narrative. Closer examination of the debate about the proposed refinery complex has yielded additional insights into the ways in which the project could have impacted the Seacoast of New Hampshire and beyond. There were other seacoast communities engaged in the fight and I hope future scholarship will add their insights and voices to the historical record. Regrettably, many of the people who were part of this history are no longer with us, but their legacy lives on. This essay is dedicated to all the people who had the passion and courage to protect and preserve the Seacoast community for future generations. They took on the mighty and the powerful in the "battle of the century."

Acknowledgments

F OR THIS ESSAY I gratefully acknowledge the following people for their guidance, support, insights, and generosity. Kurk Dorsey, professor of history at the University of New Hampshire, is a gifted scholar and writer. His keen intellect and generous spirit guided my research from the start, and course corrected my path whenever I went "down the rabbit hole" to another place and time. Catherine Peebles, my adviser and a brilliant scholar of literature, helped advance my research right through publication. Alex Herlihy, president of the Rye Historical Society, supported my research at every turn, gave many hours of his time so I could access the Town's historical records, and shepherded this publication from concept to print. The Town of Rye is fortunate to have him as steward of its rich history. Elise Daniel, a fellow graduate student at the University of New Hampshire also researching this historical event, graciously offered to work together and interview some of the research participants jointly; this was not only productive but a whole lot of fun.

Dusty boxes of yellow newspaper clippings and fading photographs were brought to life by talking with people who lived through the history. Their stories offer richness and authenticity to otherwise static text. I hope the reader will hear their voices throughout this essay. I am grateful to Peter Horne, Mel Low, Dudley Dudley, the late Phyllis Bennett, and Nancy Sandberg for generously sharing their memories and insights. Sadly, Phyllis Bennett passed away July 21, 2015, four months after I had the privilege of spending time with her at her home along with fellow UNH graduate student, Elise Daniel. Phyllis was publisher of the start-up community newspaper, *Publick Occurrences*, which broke the story that Olympic Oil was planning an oil refinery complex for the Seacoast of New Hampshire. Phyllis led a relentless effort to inform and connect the Seacoast community with facts, and bring the truth of the proposed oil refinery complex out of the shadows of the governor's office and into the light for all to see. She taught us all what it means to be "a doer."

I am particularly indebted to Peter Horne, whom I had the great pleasure of meeting after I uncovered his name in old newspaper clippings and handwritten notes. Peter graciously shared his memories and expanded my research in important ways. The "Community Reflection" in the Appendix adds insights and memories Peter developed with Rye friends engaged in the battle against Olympic Oil. I learned from Peter what it means to be gracious, to cheer on others and really mean it, and to never, ever give up.

My husband, Bill, and twin daughters, Amelia and Faith, are the joy of my life. They encouraged my research from the start and kept me going with hugs and laughter. Their love and support fuel every word of this essay.

Narrowing In (Fly Against Fate)

Sit-
In the deep soft chair
Stare at the light outside-
Outside the window- outside
Life - outside self.

Eyelids close slowly narrowing
The light circle - the life circle
Slowly, slowly closing in
An easy enclosing of the world.

And then -
Across the narrowed view
A living crimson streak
a cardinal bird flies by.

Eyelids fly too - wide open
To the window -
The whole world's out there still
For the taking. Don't sit - don't walk
Fly

Rye and Beyond: The Poetry of Jessie Haig Herlihy
Reprinted with permission.

Introduction

THE EARLY 1970s was a troubled time in the United States. The economy was fumbling after two decades of prosperity. Unemployment was rising and so was inflation. Americans were tired of the Vietnam War and distrustful of government. Americans had suffered through too many lies and squandered blood and treasure in an unpopular war. In October 1973, things went from bad to worse. The United States experienced a shock to its oil supply. Arab countries refused to sell any oil to the U.S. in retaliation for its support of Israel in the Yom Kippur War. Crude oil prices soared, home heating oil deliveries were delayed, and gasoline at the pump was harder to come by. President Nixon urged Americans to conserve energy by turning down their thermostats, driving no more than 50 miles per hour, and eliminating any long distance driving unless absolutely necessary. Gas stations were ordered to close on weekends from 9 p.m. Saturday to midnight Sunday. The president unveiled "Project Independence," a plan designed to reduce Americans' dependence on foreign sources of energy by 1980. Americans were encouraged to develop all available resources with existing and new technologies, including nuclear energy, and to reduce demand by eliminating non-essential uses of energy.[1]

In January 1973, Meldrim Thomson began the first of three terms as governor of New Hampshire. At the start, he began advocating for an oil refinery and nuclear power plant in New Hampshire. He had big political aspirations and hoped to earn the favor of the Republican Party. The good governor directed his administration to develop all state resources for the greatest good of its citizens.

The Seacoast of New Hampshire was in for the fight of its life to protect its limited coastline against the governor's plans for development. One of the world's wealthiest men, Aristotle Onassis, answered the governor's invitation with a proposal from his company, Olympic Oil, to construct an oil marine terminal, pipeline and refinery complex that would be the largest ever built from the ground-up in the United States.

This essay examines the debate about Olympic Oil's proposed oil refinery, offshore supertanker terminal, and pipelines for the Seacoast region of New Hampshire in the fall of 1973. This event is an important part of New Hampshire's history in which politics, the economy, and the environment intersected. To date, studies have largely focused on the town of Durham, and its citizens' response to siting an oil refinery along the Great Bay estuary. This essay will "widen the lens" and examine another town embroiled in the debate—the town of Rye—thus expanding the narrative of this history and giving it greater dimension. Sharper focus will be given to the offshore supertanker terminal and pipeline complex within the refinery project—elements which could have completely transformed the state of New Hampshire's coastline and economy.

The study of this event will be situated within the broader context of the 1973 energy crisis in the United

States, thus deepening our understanding of the ways in which support and opposition were shaped by the culture, economy and politics of the time.

In this essay, I argue that Olympic Oil's refinery project was made up of interdependent, yet distinct parts, each needing the other to make the overall project work, and each presenting its own set of challenges.

The supertanker marine terminal proposed by Olympic Oil would have enabled very large crude carriers (VLCCs), so called supertankers, to berth along the Isles of Shoals—a cluster of offshore islands located about nine miles off the town of Rye's coastline in southeastern New Hampshire. Large amounts of crude oil would have been offloaded and distributed by pipeline for refining inland. The supertanker terminal and pipelines would also have supported the *transshipment* of crude and refined products back out to other New England states and beyond.

The oil refinery in Durham Point would have been the initial center of operation, however, the proposed supertanker marine terminal and pipeline complex would have provided the infrastructure to support the potential development of *multiple* refineries, as well as the development of "satellite" industries—e.g., plastics, paint, and solvent manufacturers—which cluster around refineries to optimize their supply of petroleum products. Access to a supertanker marine terminal and pipeline complex would have turned the waters off Rye's coastline into a shipping gateway to all of New England and beyond.

The supertanker marine terminal would have impacted shipping traffic lanes all along the coastline, magnifying the risk of tanker collisions, as well as increasing the potential for an oil spill. The supertanker marine terminal and pipelines would have formed the infrastructure needed to transform New Hampshire into an industrial state, posing an even greater cultural and environmental threat than the initial refinery proposed for Durham Point.[2]

Prologue

T HE JUNIOR HIGH SCHOOL is typically quiet in the evening. The halls settle down after a long day of activity, waiting for the children to return in the morning. This night would be different. On January 23, 1974, in the small coastal community of Rye, New Hampshire, nearly a thousand residents came out in the bitter cold to pack the gymnasium beyond capacity. The crowd was an eclectic mix—new families with young children sat down next to town elders, with deep ancestral roots, and streets that bear their names. The well-heeled set from Rye Beach stood alongside the lobstermen of Rye Harbor, who earn a living out of these waters, and don't take kindly to jawbone speeches. Tonight they all wanted one thing—answers. The media had been covering the story for months, trying to nail down details. An oil refinery was coming to New Hampshire, and it was going to be a $600 million project. Aristotle Onassis, the Greek shipping tycoon, was going to build it— right here through the heart of New Hampshire's seacoast.[3]

All Roads Lead to Olympic...

I N EARLY NOVEMBER 1973, *Publick Occurrences*, a start-up
community newspaper based in Newmarket, New
Hampshire, was the first to run the headline "Options
on Durham Point for over 1000 acres," with illustrated
puzzle pieces for options on land being mysteriously bought
up by out-of-town realtors claiming they represented everything from a game preserve owner to a gentleman farmer.
By mid-November, news of similar efforts to buy up land
throughout the town of Rye began to spread, and residents
suspected it was somehow connected to the Durham land
deals and the governor's plans. Some Rye residents were
being approached repeatedly to grant land options. One local
newspaper reported the refinery would be "fed by a pipeline from a floating harbor somewhere off New Hampshire's
eighteen-mile coastline."[4] Residents throughout the Seacoast
followed the governor's statements looking for clues, and on
November 27, 1973, listened intently as he announced that
Olympic Oil, owned by Aristotle Onassis, was planning to
build an oil refinery in New Hampshire.

The refinery was going to be a triumph for New Hampshire and jet fuel for Thomson's political career in the
Republican Party. William Simon, the head of the Federal

Energy Administration, reportedly took notice and would later personally congratulate the governor by telegram for his efforts to support President Nixon's energy plan. The authenticity of this telegram has been disputed. Simon's message read:

> Congratulations on the progress you are making in New Hampshire toward the construction of a refinery in your state. This forward step and your progress with the proposed nuclear generating station will substantially help our country achieve energy independence by 1980 in accord with the president's goals. You may be assured that when your refinery becomes operational, the additional availability of heating oil and gasoline will be of substantial benefit to the citizens of your state and region.[5]

The fact that crude oil was to be sourced exclusively from foreign countries seems to have been lost in translation. William Loeb, publisher of the *Manchester Union Leader*, and Thomson supporter from early on, vowed to do all he could to make the governor's plans succeed. Loeb had a powerful role to play as the publisher of the only statewide newspaper in New Hampshire. He intended to use the editorial pages to win hearts and minds, and excoriate any opposition. A refinery was needed by New Hampshire to control its supply of energy and Aristotle Onassis was going to build it, bringing good jobs and tax revenue to the state. The governor emphasized it would be bigger than any refinery deal Maine was reportedly working on with Gibbs Oil in Sanford, and bigger than any refinery complex yet built from the ground-up in the United States.

Numb. — 12

PUBLICK
OCCURRENCE

November 2, 1973   *The New Hampshire Seacoast Community*   cents

Newmarket battles Oyster River	Theatre-By-The-Sea's opener	The local drive to get Nixon	Lega... over...

Options on Durham Point for over 1000 acres

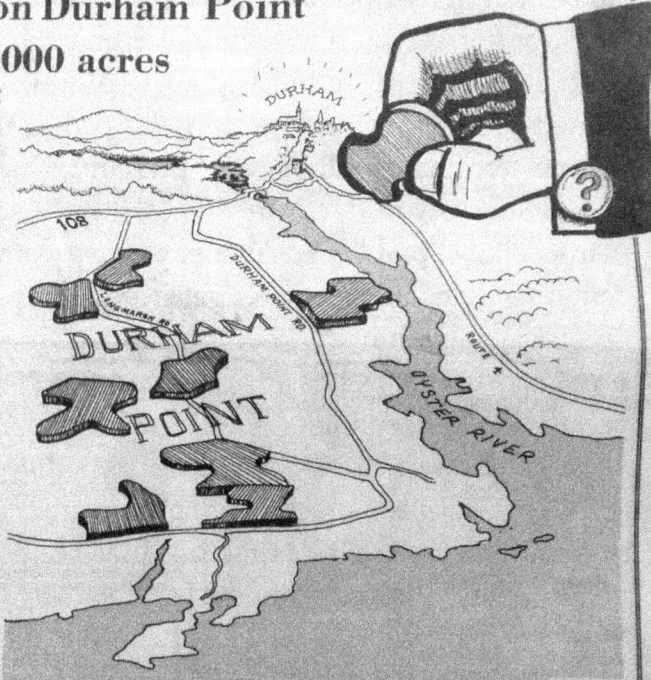

BOB NILSON

Eleven Durham Point landowners acknowledged this week that they have granted a total of over 1,000 acres in land options to a Nashua real estate dealer, George Pappademas.

In a telephone survey by *Publick Occurrences*, it was found that Pappademas is making considerable headway in his effort to sew up options on the Point. However, a frequently cited figure of 2,000 acres in options could not be confirmed.

Pappademas is concentrating his activity on the interior of Durham Point, and showing only a casual interest in shore frontage. According to landowners, he is attempting to put together a pattern of purchases. While he seems unconcerned whether the land is swampy, ledgy, open or wooded, he is trying to purchase parcels of land which fit together into a large block.

Pappademas, who, according to Lane Staude of the Nashua Board of Realtors, has been a well-known businessman in Nashua for years—and who has held a real estate license for something over a year—has been active on Durham Point for over 2 months.

Although Pappademas declined to discuss his plans this week, he has indicated to some Durham Point residents that he will make a public statement in the near future.

None of the many people interviewed were able to shed light on the use—if any—to which the land will be put. Nor was anyone able to say with certainty how many options will actually be exercised. Pappademas himself said in an October 5 interview that he had obtained seven options. He said the land would be used for his own investment, for a game preserve for friends in Keene and possibly a family compound.

Views expressed this week on the matter ranged from indignation to concern to wonderment to amusement. Alden Winn, chairman of the Durham Board of Selectmen, is watching the Point closely.

Other town officials, including Selectmen Owen Durgin and Lawrence O'Connell and Town Engineer Henry LeClair, adopted a wait-and-see attitude.

Winn is confident that the Durham zoning ordinances are a strong safeguard, but said that anyone determined to challenge them could tie up the town in a drawn-out court action.

Rebecca Frost, zoning board member, is jittery because Governor Meldrim Thomson has recently suggested that town zoning ordinances be waived to permit critical industries to enter the state.

In the October 5 interview, Pappademas said that he fully intended to observe the Durham zoning ordinances.

Some observers suggested that a state statute might force disclosure—at least of the buyers. However, the law in question, RSA 356-a, does not apply in this instance, according to Frederick Upton, a Concord lawyer with much experience in law involving real estate.

Minnie Murray starts drive

Durham petitioners want Nixon out

For two hours a day or so, a Durham woman has been standing on a business street collecting signatures of citizens who want to see the President of the United States impeached. For a couple of days the signatures were being collected at the rate of 300 an hour.

On Monday and Tuesday of this week, they were still being written at a rate of 50 an hour.

To date, more than 600 signatures have been collected all to be sent eventually to keep the pressure on Congress to instigate the presidential trial procedure.

Minnie Mae Murray of Durham first showed her activist nature in public when she circulated a petition in the Pentagon protesting the food in the cafeteria.

That was back in the '40s, she said, when she was working as a secretary in the old War Department. She even succeeded in getting the signature of Secretary of War Henry Stimson's secretary.

Particularly noisome, she recalled, were the stewed tomatoes, which were "faintly pink and composed mostly of filler."

"We got some real response to the petition," she said. "Once it started around it snowballed and we got some action."

That's just what she hopes will happen to her latest petition drive—asking for the impeachment of President Nixon.

Mrs. Murray, 53, is the wife of Pulitzer Prize-winning writer Donald Murray, a professor of English at UNH.

She is a mother of three who describes herself as an activist, but "not political."

She began the petition drive two weeks ago because, she said, "I couldn't sit still any longer." Candid as crusaders go, she says her dislike of Mr. Nixon goes back a long way— "I'm certainly not open-minded," she said last week.

She claims that, although she's a life-long Democrat, her motivations aren't partisan. "If Nixon had been a Democrat, I'd have done the same thing."

While she has thought Mr. Nixon should be impeached for some time, it wasn't until after the so-called "Saturday night massacre" that she felt compelled to try and help bring it about. The "massacre" was the day (Oct. 20)

Turn to IMPEACH, page 12

Olympic released the first of a series of reports for the project in November 1973. The refinery would cost approximately $600 million to construct. Two thousand five hundred people would be employed for about two years; a thousand thereafter. The refinery would process 400,000 barrels of Arabian crude oil per day. Crude oil would be unloaded at the supertanker marine terminal along the Isles of Shoals for distribution to refinery plants (plural). Crude oil and refined products would be transshipped at the supertanker marine terminal to other New England states and beyond. The refinery would "form the core necessary to develop petrochemical plants and further downstream processing. ... Tentative plans include such petrochemical plants."[6]

In late December 1973, Aristotle Onassis toured the proposed refinery site, at an altitude of 20,000 feet in his private jet. He then joined the governor and Olympic's hired consultants for a press conference in Bedford, New Hampshire. It was a sleek production, but the question and answer session left Rye residents feeling that the governor's statements were long on promises, and short on details. What did Olympic's representatives mean by "Exact uses of the island have not yet been established" when questioned about the Isles of Shoals? And how did the supertanker marine terminal and pipelines fit within the refinery planned for Durham? Was all of this connected to the obscure real estate activity over the past several months?[7]

The town of Rye needed more than speeches. It organized an information session to get a better understanding of exactly what was being planned for Rye and its coastline. Representatives of Olympic Oil and their hired consultants were on the panel. They were joined by Rye town officials, Richard Keppler, an offshore terminal expert from the Environmental Protection Agency, Frederick Hochgraf, associate professor of materials science and metallurgy at the University of New Hampshire, to address the pipeline risks, and Dr. John Kingsbury, director of the Shoals Marine Laboratory, to address the potential impact on marine life around the Isles of Shoals. At 7:30 in the evening on January 23, 1974, nearly a thousand residents from every part of town packed the junior high school. The information session lasted over three and half hours, but the defining moment of the evening came when Olympic Oil showed a film of a Kuwait oil company constructing a tanker marine terminal with underwater pipelines. The film was narrated by a British voice, and had the soundtrack of a foreign country's anthem playing in the background. The event was widely covered by the media. *The New York Times* described the film this way: (It) "showed an underwater dynamite charge

"In the path of the pipeline - Rye waits." Publick Occurrences, *December 7, 1973.*

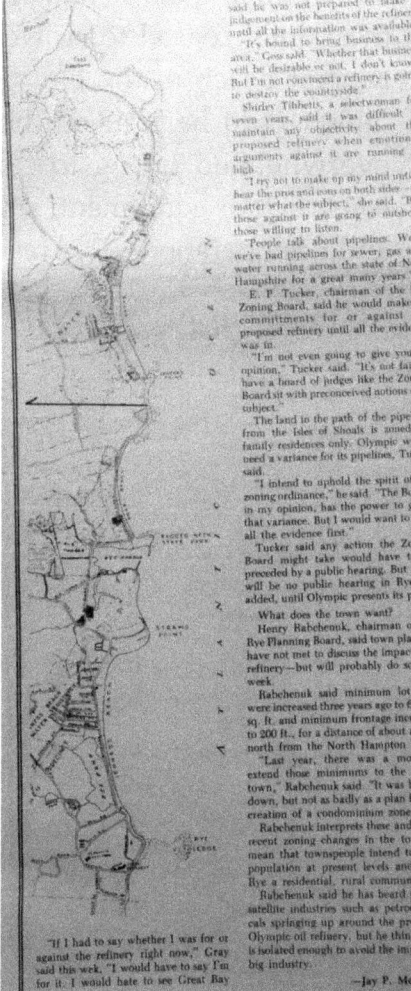

blowing up what was described as 'obstinate rock' and the beach being dug up for huge, concrete-encased pipes. . . . After being told that this would not happen on Rye Beach, Ernest E. Tucker, the chairman of the Board of Adjustment, grumbled, to applause, 'I'm not so sure of that.'" Lobstermen were horrified as they envisioned the seabed being blasted by dynamite, and their livelihood working the sea with it. Motel and restaurant owners tried to imagine why any tourist would spend their vacation dollars along a blown-up and blackened coastline. And homeowners feared their quiet coastal town would be blown up like the sandy Kuwait beach in the film. Rye's tourist and fishing economies would be dead in the water and the town would be overwhelmed trying to cope with transient workers and construction equipment, stressing the roads, schools, police and fire operations. Rye residents were speechless.[8]

Numb. — 23

PUBLICK
OCCURRENC ;

January 25, 1974

The New Hampshire Seacoast Community   cen

A legislator on UNH Gays

The Gingerb___ ___ Lady

The Seabrook compromise   reviewed

Few Olympic facts for Rye

If a resounding ovation means anything, Pete Tucker summed up the mood of the standing-room-only audience assembled in Rye Wednesday night to hear about the proposed Olympic oil terminal and pipeline.

"There have been a lot of questions, but no answers," Tucker told Eugene Harlow, the F. B. Harris Company engineer charged with designing a deep-water oil terminal for Olympic Refineries, Inc.

Tucker, an engineer and chairman of the Rye zoning board of adjustment, grilled Harlow on the question of whether it would be necessary to prefabricate structures on the beaches of Rye before assembly of an off-shore terminal, at the Isles of Shoals.

Tucker and the other 800-odd people packed into the Rye Junior High School gymnasium had just watched a film presented by Harlow, depicting construction of a fixed-berth terminal in Kuwait. At one point in the film massive towers were pushed across a beach to the water by bulldozers.

Harlow insisted that "the kind of prefabrication you saw in the movie will not occur on your beach." After a heavy moment's pause, Tucker replied, "I'm not so sure of that."

Tucker was unable to get any answers about whether the town would have control over use of its beaches by the terminal construction crews.

Visibly exasperated himself, Harlow replied to accusations of evasiveness by saying that to expect him to have specific answers before the completion of his feasibility study was like "asking a man in the middle of a race if he is going to win."

Douglas Gray, chairman of the Rye board of selectmen, retorted, "Why

Turn to RYE, page 2

Lobstering in New Hampshire, the hardships and the worth of it. (Pages 10 and 11)   —photo by George Burke

84% in poll oppose refinery

A random sampling of opinion among 121 registered voters in Durham conducted for *Publick Occurrences* last week reveals that 84 per cent are opposed to the location of an oil refinery on Durham Point.

The poll, designed to allow respondents to express slight, moderate or strong feelings on the refinery, showed 10 per cent in favor—and six per cent with no position at all.

The poll, conducted between January 16 and January 22, also attempted to determine whether the bulk of opposition is confined to University of New Hampshire families, or spread across the entire community.

The results: 94 per cent of university families are opposed, four per cent are in favor, and two per cent have no position. Among non-university families, 73 per cent are opposed, 17 per cent

are in favor, while 10 per cent take no position.

"It appears that, contrary to much speculation, University-employed families do not at this time constitute the swing vote on the issue," the poll said. "Non-university families also demonstrate considerable opposition to the refinery."

Turn to POLL, page 2

Seabrook plant compromise

"Few Olympic Facts for Rye." Publick Occurrences, January 25, 1974.

2

The People Mobilize...

I hold it that a little rebellion every now and then is a good thing.

—Thomas Jefferson

BEFORE THE JANUARY 1974 information session, Rye townspeople had an open mind about the refinery project, preferring to gather more information before coming to any conclusions.[9] The scale and scope of the refinery project, however, was beginning to sink in, and it created a sea change in residents' attitudes—they began to mobilize. Concerned Citizens of Rye (CCR) was formed, chaired by Peter Horne. It united the opposition of Rye residents and embarked on a relentless campaign to stop the Olympic Oil project. Town meetings, newsletters, letter writing campaigns to state and federal legislators, community fundraising events, attendance at legislative sessions, etc. were all organized to raise awareness and defeat Olympic. Rye's centuries-old heritage, semi-rural coastal character, and the livelihood of its fishermen, lobstermen, motel owners and restaurateurs, were directly threatened by the Olympic project, and CCR was determined to stop it.[10] CCR's newsletter, *The Coast Watcher*, was widely distributed in town, as well as to friends in surrounding Seacoast communities who had their own opposition efforts underway. CCR's official policy statement read:

> Public policies affecting Rye's future should be determined at the local level and in concert with surrounding communities.
>
> We recognize the need for sources of energy and urge that the people of New Hampshire support intelligent, coordinated New Englandwide planning to meet the energy needs of New Hampshire and the New England region.
>
> We subscribe to the United States Environmental Protection Agency's recommendations that heavy industrial oil refineries, pipelines and terminals should be located away from sensitive salt water estuaries and coastal areas.
>
> We oppose the stated plans of Olympic, Inc., for a refinery and offshore terminal system and will oppose any State of New Hampshire proposal which presents the same dangers to the quality of life in Rye and the Seacoast area.
>
> We will take whatever action is appropriate in coordination with other area organizations of like mind to carry out this policy.[11]

CCR's statement was carefully crafted for a reason—it was putting Olympic *and* the Thomson administration on notice. Early on, the governor made it clear that the coastline of New Hampshire was a resource, and as such, was to be utilized for the greater good of *all* citizens. George Gilman, the commissioner of the Department of Resources and Economic Development (DRED), at the request of the governor, studied developing an oil refinery and deepwater marine terminal in the state. The so-called DRED report advocated public ownership of a deepwater marine terminal in southern New Hampshire, among other things. According to the report, the state could collect millions of dollars in tax revenue by levying so-called throughput fees—taxes assessed on every barrel of oil coming through the terminal.[12] The public versus private ownership debate prompted CCR to cover all conceivable forms of development.

Many people throughout the Seacoast worked day and night to stay on top of it all and keep the pressure on Olympic. The largest opposition group on the Seacoast was Durham-based Save Our Shores (SOS), chaired by Nancy Sandberg. Rye's largest organized group was CCR; it was chaired by Peter Horne and it coordinated with SOS, particularly with regard to technical assistance from University of New Hampshire specialists. Many Rye residents were dedicated to the fight; among them were: Jessie Herlihy, Louise and Charles Tallman, Guy and Madeline Chichester, Frances Holway, Peter and Holly Horne, Mike and Sally Flanigan, Bill and Priscilla Jenness, Mel and Jean Low, Doug Gray, Betsy and Bill Bischoff, Larry Schell, Sharon Fish, Judy and David Paterson, Gail Tucker Hughes, Marion and Tom Barron, Frances Tucker, Bernice Remick, and Snick Wilson.

Mrs. Louise Tallman was chair of the Rye Conservation Commission at the time of the refinery controversy. On January 24, 1974, one day after Olympic made its presentation to Rye residents, she wrote a letter to Malcolm "Tink" Taylor, the executive secretary of the New Hampshire Association of Conservation Commissions. Mrs. Tallman expressed her frustration over the ecological damage that would be sustained by land in the path of the pipeline and refinery. Her letter captures the sentiment of many Rye residents.

Mrs. Tallman wrote:

> I feel that there should be a description of
> the ecological losses of the land area of pipe-
> line and refinery. Each Conservation Commis-
> sion could go a long way on this in each town
> concerned. Our statements might come out
> a bit different from that hired by Olympic.
> Besides the Mill Field…the line would pass
> through a fragment of climax forest, magnif-
> icent Beech trees of the same owner. These
> directly adjoin our Conservation acres. You
> can't put back 100-year old Beech Trees.[13]

Along the pipeline's path from Concord Point in Rye,
one property was particularly critical. Two elderly sisters,
Frances Tucker and Bernice Remick, owned a forty-two acre
farm located along Brackett Road, just up from Concord
Point. The sisters were pressured repeatedly to sell by
multiple out-of-town realtors working on behalf of Olympic.
The sisters were initially offered $90,000 as early as October
1973. By February 1974, the offer had increased to $200,000.
The Remick/Tucker property had been owned by the family
since the early nineteenth century and consisted of a large
farm house, environmentally sensitive wetland, and a historic
graveyard marking the 1691 Brackett Massacre where the
earliest settlers died in a vicious clash with Native peoples.
Olympic needed the sisters' property to run their pipeline up
from Concord Point through Rye to Portsmouth and on to
Durham Point. Failure to secure the Brackett Road property
would have required extensive repositioning of the pipelines
and pumping stations planned along the route. At one of the
last attempts to buy the property, the sisters refused to even

Two ladies of Rye thwart pipeline plans

Two elderly Rye women are blocking Olympic's effort to complete a pipeline right-of-way from the coast to Durham Point.

Frances Tucker and her sister Bernice Remick have refused repeated offers to buy their 55-acre farm on Brackett Road, which is directly in the path of the proposed pipeline.

Mrs. Tucker and Miss Remick refused the latest Olympic bid last week when Jeff Marple of Marple Associates, Portsmouth, working on Olympic's behalf, offered $200,000 for the property.

The offer was made by telephone, since the two sisters have long since stopped admitting real estate agents into their house.

After successive failures early last fall to persuade the women to sell, Olympic attempted to enlist the aid of Douglas R. Gray, chairman of the Rye Board of Selectmen, Reverend Richard L. Schlafer, minister of the Bethany Congregational Church in Rye, and Mrs. Tucker's daughter, who lives in Virginia.

Gray, who is a lawyer, acknowledged that he consented to talk to the women on behalf of a client, whom he understood to be representing a private individual. Gray also made a phone call to Mrs. Tucker's daughter, to ask her help, but she refused to cooperate.

Turn to PIPELINE page 2

Two ladies of Rye thwart pipeline plans. Publick Occurrences, *March 1, 1974.*

let the realtor into the house, but noted in one newspaper account, "He had already taken off his rubbers." The sisters Tucker and Remick did ultimately sell their land—to the Rye Conservation Commission—for $12,000. It remains protected open space.[14]

Star Island and Lunging Island—two islands in the Isles of Shoals under Rye's jurisdiction—were also at the center of controversy. Olympic planned to situate two fixed-berth supertanker terminals along the islands for unloading and loading operations. Olympic planned to use Star Island for positioning of navigational equipment, and Lunging Island for a pumping station.

"I thought to myself—they just can't have it"

...PIPELINE from page 1

Gray said in an interview that he became suspicious that his client was attempting to acquire a pipeline right of way, and that he immediately withdrew his services because of possible conflict of interest.

The two women told *Publick Occur-*

The offer was made by telephone since the two sisters have long since stopped admitting real estate agents into the house.

rences this week that they have no intention of selling.

Mrs. Tucker, who acted as spokesman, put it simply: "I thought to myself—they just can't have it!" Miss Remick readily seconded her sister.

Mrs. Tucker and Miss Remick live with an elder sister, Mary Watson (who does not share ownership), in a large, white farm house on property which has been in their family since the early nineteenth century.

"We're old ladies," said Mrs. Tucker. "We think they're trying to take advantage of three old ladies, and they're not going to get away with it."

Bernice and Mary are both in ill health.

The sisters were first offered $90,000 in mid-October by Emile Soucy, a Hampton realtor. They were also offered life tenancy in the house. Mrs. Tucker said that she and Miss Remick refused the offer outright. Soucy returned for a second and a third try. On the third occasion, he was refused admittance to the house.

Soucy then approached Douglas Gray for help. Although Gray would not confirm that Soucy had come to him, he acknowledged that his client was a real estate agent.

Gray said that the real estate agent, Soucy, told him, "I represent someone in Rye who wants to buy land from these women." Gray agreed to visit the women, and called on them twice, but to no avail.

Gray said that he then received a telephone call from the individual represented by Soucy. (Although Gray would not reveal the person's name, *Publick Occurrences* has learned that it was Peter Booras, the organizer of Olympic's land acquisition effort.)

Booras told Gray to call Mrs. Tucker's daughter in Virginia. Gray did so, with no success.

By early November, Gray knew enough about land options being obtained in Rye to suspect a pipeline. In a mid-November phone conversation with

Booras, he pulled out of the affair, saying that as a selectman, he could not risk a conflict of interest.

Immediately thereafter, he told the other two Rye selectmen of his involvement and of his decision to pull out. This was about two weeks before the refinery announcement by Governor Thomson.

On November 30, after the Governor's announcement had confirmed his suspicions, Gray wrote a letter to Booras, saying he could not be involved in the matter. He said that he billed his client for services ($100), and has been paid. He said that he has had no contact with Olympic real estate people since then.

"You can't serve two masters," Gray said, "It would be improper and unethical for me to continue to work for that client."

He added that he believed the Olympic agents did not know he was a Rye selectman, although he is sure they knew that he had had previous legal contact with the sisters.

Last year Gray drew up a deed on land purchased from the sisters by the Rye conservation commission. They were willing to sell 42.5 acres to the commission for $12,000, even though they have since turned down $200,000 from Olympic for 55 acres.

Mrs. Tucker said that she and her sister were approached again "about a month ago," this time by George

Pappademas, the Nashua realtor. Mrs. Tucker said she refused to let him in the house, even though he "had already taken off his rubbers."

Olympic tried again last week through Marple Associates, and raised the offer to $200,000.

Speaking of the offer, Mrs. Tucker exclaimed: "What would we do with all that money!" Asked what would happen if they were offered $1 million, she laughed and said, "That would be even worse! That money doesn't bother me a bit." She also expressed concern over the fate of the fishing industry if the project goes through.

The sisters are fiercely independent. Asked if she drives a car to do errands, Mrs. Tucker replied: "I should hope so. We'd die of starvation otherwise!"

She and Miss Remick tend a small garden and work in the yard a lot. Although their father died in 1931, they still list their telephone number under his name. Since their mother's death in the early '50's when she was over 90 years old, the sisters have owned the farm.

They live quietly in Victorian surroundings: An ornately carved settee, originally upholstered with horse hair, cuts a diagonal across a corner of the parlor. Knick-knacks pack shelves and window sills. Furtive cats peer out of dark corners. The walls are covered by aging wallpaper with elegant floral motifs.

Old family portraits cluster on top of a vintage piano, and shelves of books stand at arm's length from comfortable wing-back chairs. In this room, the television set with a TV guide on top is a startling discovery.

Outside, a long-abandoned orchard arches in a confusion of limbs across a stonewall and over the road.

The property line begins in lowland below the house, rises, continues for a quarter of a mile along the road, then ends in another low marsh. A pipeline on either side of this property would have to pass through wetland.

Out back gray birches and sumac invade abandoned hay fields. But the towering barn, like the sharp-featured house, remains straight and proud. These buildings seem to echo the defiant challenge of Mrs. Tucker: "If I had a chance to tell Onassis what I thought, I'd tell him to go back to Greece!"

* * * *

This week, the two sisters composed an open letter explaining their feelings which was read by a representative of the Concerned Citizens of Rye to the state legislature.

—Ron Lewis

1. The land Olympic wants

2. Land sold to the town

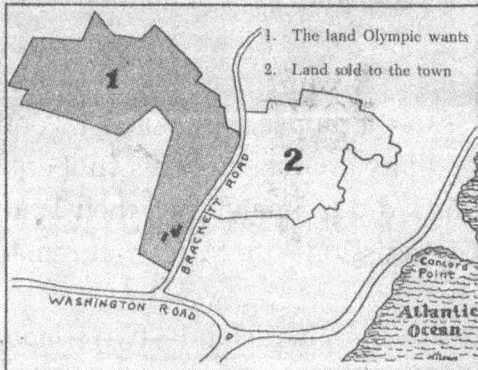

Sister Soldiers. Frances Tucker and Bernice Remick. Publick Occurrences, *March 1, 1974.*

The Randall family had owned Lunging Island for more than forty-eight years and had deep emotional attachment to the land, which included the burial ground of one of their sons. Beginning in late November 1973, Olympic representatives began calling on the Randalls, increasing the pressure and changing the terms of sale with each visit. Robert and Prudence Randall were not only pressured, they were misled. Originally, Olympic's representative tried to persuade the Randalls to sell the island for development as a resort. After direct questioning by the Randalls, the Olympic representative admitted the option was being sought in connection with the oil refinery project. The Olympic realtor also misled the Randalls into believing other islands in the Isles of Shoals had already sold out to Olympic. Distraught, the Randalls contacted five different lawyers, and were advised their property could potentially be at risk if the state wanted to take the land by eminent domain for the greater good of the public. In the event Lunging Island was taken over by the state, the Randalls would receive only a fraction of the offer on the table from Olympic. Mistakenly believing that other Isles of Shoals owners had already sold out, and fearing they might lose their land to the state, the Randalls optioned their land to Olympic on December 7, 1973 for a reported, but unconfirmed, sum of $500,000. It was the fifth and final visit by Olympic. The Randall family was devastated.[15]

Star Island Corporation owned Star Island and Appledore Island, one of Maine's islands that houses the Shoals Marine Laboratory. It issued a public statement to Olympic that said:

> We cannot conceive of a set of circumstances, financial or other, which would induce us to alienate Star Island. We believe we have

Mrs. Randall, on giving up her island:

'It's like a death in the family, of somebody very close'

LUNGING... from page 1

19th century writer and artist Celia Thaxter, who lived on neighboring Appledore after spending her honeymoon in it. The original building, built about 200 years ago, was rebuilt about 100 years later by Oscar Laighton, Celia's brother and a Shoals historian, Mrs. Randall said.

The Randalls are steeped in the lore and tradition of the place and it's obviously in their blood. "It's like a death in the family of somebody very close," Mrs. Randall said of having to give up the island. "You have to face it but still you can't accept it."

In the living room of another historic home in Danvers, Mass., last Sunday, Randall told the story of his encounters with Olympic representatives.

(Their Danvers home, called the Prince-Osborne house locally, was originally constructed in 1634 while the town was still part of Salem Village, and is one of the very few early 17th century houses still standing in the country. It is the home of Goody Osborne in Arthur Miller's play "The Crucible," about the Salem witch trials.)

In the dimly lit room, spare of furniture against the back-drop of ancient dark brown floor-to-ceiling panelling surrounding a low fireplace, Mrs. Randall sat in an old country Windsor chair by the only table while the men, Robert and son Ray, alternately stood and paced about.

The first encounter with refinery people, Randall said, was in late November when Booras contacted Ray in Manchester at the home of a friend.

"Booras chased him clear up to Manchester, New Hampshire, where he was visiting a friend, and said he had to see him right away."

Mrs. Randall said that Booras had traced her son through Rye fishermen, having learned that the tax bill for the island property is in her name, and apparently assuming that Ray was either the primary owner or foremost heir to the property.

Ray said he met Booras in a gas station parking lot and that Booras offered him "a substantial sum" for the island.

"I told him that I wouldn't sell it for a million dollars," Ray said, but told Booras that he wasn't the controlling owner anyway. "He seemed to lose interest in me when I told him that," Ray recalled.

Booras showed up at the Randall's house later the next day, having left Ray in Manchester, and made them an offer for an amount somewhat less—they later realized—than he'd offered their son.

The Randalls declined to disclose the amount of the offers on the advice of their lawyer. (Olympic agents claim the Randalls accepted a $500,000 offer.)

Mrs. Randall said Booras told them he wanted the island for "a resort development."

"A resort just didn't make sense," Mrs. Randall said. "There's only about two and a half acres of topsoil on that island at high tide (the area is about five acres) and there's only one mooring site where there's sufficient protection."

"At dead low water you can barely float a skiff into our inner harbor," she said. "How could they use it for anything like a resort?"

But, Booras insisted that was what he wanted the island for, and the Randalls told him they weren't interested in selling—for any reason or at any price, the Randalls said.

When they heard nothing more about the matter for a couple of weeks, Randall said, they thought the whole thing had blown over until they were visited by one George Stamatelos of Keene, a real-estate man who said he was representing Booras.

In the intervening period, the Randalls had read press accounts of the Onassis oil refinery proposal for the New Hampshire seacoast, and Robert said he began to put two and two together. The vast amount of money being discussed and the apparent urgency and secretiveness of Booras plus Olympic's plans for a tanker terminal near the Isles of Shoals could only add up to one thing.

Stamatelos came with his son-in-law David Mann, also a Keene real-estate broker, to repeat Booras's offer and discuss the matter more fully with the Randalls. Randall asked him directly if the sale was related to the oil-refinery project.

Stamatelos first claimed he didn't know what Booras wanted the island for, Randall said, then after repeated queries, admitted it was to be part of the refinery project.

Stamatelos is president of Interstate Realty of Keene, a recently-formed company. He formerly sold insurance, according to Mrs. David Mann, his daughter who answered the phone for the agency.

Her husband was out, she said, and Stamatelos was in Florida (where they would soon join him), but she was able to provide some information, nonetheless.

Stamatelos is a good friend of Peter Booras, she said. "We all go to the Greek church here."

Her father and her husband "have been working for Mr. Onassis," she said with evident pride, adding that Stamatelos had had to operate under cover on this particular job. She said he worked at obtaining options for the company mainly in the Portsmouth area.

"His cover wasn't broken until way at the end," she declared.

According to Randall, Stamatelos told him that he had no authority to alter the conditions of the offer first made by Booras and would have to refer all questions on changes back to him. But later in the discussion, Randall said, he was promised that the family could have life tenancy on the island, as their house—at the north end—would not be in the way of the company's pumping station planned for the south end. Also, Randall said, he was told the company might consider a lease.

Being informed the family still wasn't interested in selling, Stamatelos left, Randall said, only to return the next night with a different story.

This time Stamatelos said the company wasn't interested in any lease arrangements, and could not offer them life tenancy in their cottage, because the entire island was needed for the project.

Since the company planned to tear down all the buildings, though, the Randalls were told, they could keep the cottage and guest house if they wanted to remove them from the island. Again, they said they weren't interested.

Before Stamatelos paid another call—about ten days later—Randall decided to talk to some lawyers. He said he spoke with five different corporate lawyers about the matter and they all told him that if the refinery project went through the state would likely take their land by eminent domain as being for a public purpose to provide oil in an emergency shortage, and they'd probably be paid a lot less for it than Olympic was offering. Also, there would probably be few fringe benefits, like being able to save their buildings.

They settled on one lawyer, Israel Block of Lynn, Mass., to handle their affairs in this matter, and had him draw up a revised proposal to present to the company, which included certain restrictions and protections they considered important if an agreement on the island was to be reached.

"We didn't even want to consider it," Mrs. Randall said, "but we were advised that this was the only way we could protect ourselves."

Block, reached at his home in Swampscott, Mass., would offer no details of what he advised the Randalls to do, saying this is a highly confidential matter between lawyer and client.

He said he is not familiar with New Hampshire laws, nor did he consult with anyone who is familiar with those laws before giving his advice. However, he said he assumed the laws are similar to those of Massachusetts, in that Common Law and statutes do not bind the state to local zoning or other restrictions and allow it to take land whenever it wants for an acknowledged public purpose.

He cited an instance in which the Woods Hole, Nantucket & Marthas Vineyard, Mass. Steamship Authority managed to obtain land in Hyannisport for its use through the Massachusetts legislature by eminent domain.

The land was obtained by the state, he said, from private owners, despite their protests, and leased to the authority for use as a parking lot for their ferry service from Cape Cod to the offshore islands. The matter was fought to the state Supreme Court, he said, which ultimately ruled in favor of the state and the authority.

Block acknowledged there is a difference between having a state take land for the benefit of a quasi-public body such as the steamship authority, and taking it for a private firm like Olympic Refineries. He also didn't dispute the notion that an obvious primary concern for preserving their island such as the Randalls profess might make the risk more worth taking. But he said, "They came for my advice only after they had an option in their hands," and added that he was not present at any meetings between his clients and Olympic representatives.

When Stamatelos came again (this was the fifth and, as it turned out, last meeting between the Randalls and Olympic agents) Robert Randall was ready with Block's revised proposal.

Stamatelos looked at it, Randall said, and commented on several sections of it, saying they were "tough." One provision, that the company would pay the costs of barging the Randall's cottage off the island to another location, he balked at, Randall said, saying the company wouldn't agree to it.

This was evidently the last straw for Randall, who didn't like the whole deal from the beginning despite what the lawyers said. "I told him I wouldn't sell the island to him no matter what," Randall said.

At this point, evidently, Stamatelos decided to get tough.

According to Randall, Stamatelos told him that if they couldn't settle the matter agreeably that night that it would be all over and he wouldn't come again even if they were to call him. Randall said the Keene real estate man told him "we'll get it otherwise" if he didn't agree to deal then and there. In that event, he was warned, they'd be lucky to get half what Stamatelos was offering.

When he still demurred, Randall continued, Stamatelos added the clincher.

"He told us that they were really after all the Isles," Randall said. "He said that they already had options on two of them and that another was 'listening.'"

"I must admit that convinced me," Randall said. There appeared to be no use in holding out if all the others were going.

It was a few days later, Randall said, when he learned that Star Island Corp., owner of most of the other islands, had publicly stated that none of them were for sale and that the organization had determined to fight the refinery proposal with all the resources it could muster.

It was too late to do anything about it. He had given Stamatelos an 180-day option on his island, with provision for at least one renewal. The date was Dec. 7.

"Pearl Harbor Day," Randall noted darkly.

Mrs. Randall said she could not conceive of not having their island to go to when winter is over. She said that when Stamatelos left with their option under his arm, he said to her, "Cheer up—maybe it won't go through."

Some Rye officials have suggested there might be another reason the Randalls decided to option their island—their tax bill went up by about 15 times two years ago.

According to Rye Selectman Shirley Tibbetts, the assessment on Landing Island was $1,200 for many years prior to 1971 when the town voted to have the

ISLES OF SHOALS

— Prudence Crandall Randall

Upon the wide and lonely sea
they lie in broken line,
Out-crop of ledge, far-flung and bare,
barren, bleak, alone.
Their craggy outline, low and black
against the eastern sky
Scarce breaks the undulating flow
of moving, surging brine

Above the fringe of moss and weed
that girds the tidal reach,
The stern and rugged face of ledge,
bleached white by wind and spray,
Arise irregular in range
from cove to towering cliff,
Each rift and chasm, inlet, dike,
a wonder and delight.

Just rocks and sea and sky it seems,
and yet, each isle is crowned
With vegetation, live and green,
in strangely sweet array;
From out the tangled weed and brush
in gay abandonment
Springs forth both flower and bird alike
with naturalness and charm.

A place of solitude and dreams
these islands out at sea,
A hallowed spot where silence reigns
in quietude and peace,
Where nature speaks in gentle tones
of graciousness and joy,
Of light and warmth, of constancy,
of unity and song.

It's like a death in the family . . . Publick Occurrences, *Feb. 1, 1974.*

To aid refinery

Thomson may skirt Durham zoning, taxes

Governor Meldrim Thomson is examining ways to circumvent Durham's zoning ordinance, despite his repeated public claims that he would not force a refinery on the town against its will.

In addition, the Governor is seeking ways to decrease the local tax burden for Olympic Refineries, although the tax benefits to Durham have been put forth as one of the attractions of the proposed refinery.

Property tax breaks and the power to supercede local zoning were subjects discussed at a meeting of Olympic officials, Governor's office representatives and officials of the New Hampshire Industrial Development Authority in that agency's office on January 15, according to Fred Goode, the Governor's special assistant for refinery affairs.

"The question of whether rezoning is within the scope of the IDA's powers was discussed there (at the meeting)," Goode told Publick Occurrences this week.

Asked if the Governor has learned anything definite about IDA re-zoning power, Goode replied: "The question is still up in the air." He said that "there was a difference of opinion" among those present, concerning the power of the IDA to "industrialize an area, not previously zoned for industry."

Asked if the group discussed property tax breaks which might be available to Olympic through the IDA, Goode replied: "The Governor's office is interested in that question. Taxes may have been part of it, but I don't have my personal recollection—not in those specific terms. There were six or seven people in the room and a number of conversations were going on."

Goode added that the Governor's Office has made inquiries on the tax question, as well as the zoning question and financing, through IDA, to the Attorney General's Office. He said that these inquiries have been answered and that "their (Attorney General's Office) responses might have touched on the tax question."

(During the telephone interview, Goode sent his secretary to search for the Attorney General's replies to Governor's Office inquiries on the powers of the IDA, but she was unable to locate them. Goode said that he would reveal their content to Publick Occurrences when they are found.)

Asked if he saw any inconsistency between the Governor's actions to determine if IDA can break zoning and tax break powers, and his public statements supporting home rule and tax advantages for towns of refineries, Goode said no.

"I see no inconsistency there at all—if the IDA does have that power, suppose that some pro-refinery group such as SOR, or the IDA itself, petitioned the Governor to use that authority. The Governor would then have to assert his home-rule position in the face of that. He needs to know what influences may be brought to bear on him in the future."

Vee Kounas, the executive secretary and sole salaried official of the Industrial Development Authority, assured Publick Occurrences that the ten-member board comprised of businessmen has "never entertained the idea of obtaining land

Turn to IDA, page 13

'Honeymoon Cottage' on Lunging Island

The taking of Lunging Island

The Massachusetts couple who optioned Lunging Island to Olympic Refineries claim they did so reluctantly and only after Olympic agents told them that similar options had been secured on two other Shoals Islands.

Robert and Prudence Randall of Danvers told Publick Occurrences this week they also were convinced by Olympic representatives and their own lawyer that Lunging Island would be obtained by eminent domain if they resisted the sale.

Evelyn Browne of Durham Point complained in November that similar tactics had been used to pressure her into giving up an option.

They also said that Olympic project coordinator Peter Booras of Keene first tried to get them to sell by going through their 24-year-old son, Ray, and later, when asked what he wanted the island for, told them "a resort development."

When the Randalls found out it was really part of an oil refinery project, their own lawyer told them they'd better accept the company's offer because they'd likely lose the island anyway if the refinery deal goes through.

Mrs. Randall said her family has been coming to Lunging Island every summer for nearly 50 years. Her children have never known any other summer place. One of her sons, Richard, a veteran of Vietnam, is buried on the island.

Their island house, "Honeymoon Cottage," was given its name by famous

Turn to LUNGING, page 10

The Taking of Lunging Island, Publick Occurrences, *Feb. 1, 1974.*

the resources, human and financial, to maintain our position on Star Island, and will utilize those resources to the fullest extent to protect our interests.[16]

Such unwavering Yankee opposition confronted Olympic with a set of challenges they never did overcome. Olympic needed a supertanker terminal. No crude oil, no refinery operation. But why were the Isles of Shoals so critical to Oympic's plans? John Kingsbury, the director of the Shoals Marine Laboratory on Appledore Island, answers this question best: "To place a terminal in the lee of the Shoals is

Star Won't Sell

Dec. 15, 1973.

BOSTON (AP) — The Star Island Corp. says it has not been contacted since Olympic Refineries acquired an option to buy one of the Isles of Shoals, a spokesman said.

An Olympic Refineries proposal for a $600 million Durham Point oil refinery said the oil would be "unloaded at an offshore deepwater monobuoy in the vicinity of the Isles of Shoals for delivery to the refinery processing plants."

The non-profit organization owns three of the Isles of Shoals, two of them in New Hampshire territorial waters — Star Island and Appledore Island.

The spokesman said an Olympic representative contacted a member of the Star Island Board of Directors about three weeks ago.

"He was told the islands are not for sale under any circumstances and we haven't heard from Olympic since," the spokesman said.

The spokesman added that it is "impossoble to put a price on the islands. It has no value in dollars and cents."

Star Island is used during the summer as a conference center for educational and religious groups. Appledore has a marine biology research station operated jointly by Cornell University, the University of New Hampshire and the State University of New York.

"The only way they're going to get our islands is over our dead bodies," the spokesman said.

to bring it within the three mile limit (which is drawn around the islands as well as along the mainland coast). In fact, close to the Shoals is the only location along the New Hampshire coast where the water is deep enough for supertankers to get within the three-mile limit. Neither federal nor local governments control the ocean bottom from tide line to the three-mile limit. That three mile span is solely under control of the state."[17] Star Island's checkmate to Olympic.

"Star Won't Sell."
Portsmouth Herald,
December 15, 1973.

Oceanic Hotel, Star Island. Photo courtesy of Rye Town Museum.

3

The Devil is in the Details...

TIME AND EVENTS MOVED quickly after the governor's November 1973 announcement. Olympic released several reports, but there were big gaps in information and many moving parts. It was not until late February 1974, just ahead of the town votes in Rye and Durham, that Olympic finally released complete details of the refinery project: the type of supertanker terminal to be developed; the site of the refinery; the amount of water and energy needed; the path of the pipeline complex; environmental safeguards; truck terminals; pumping stations; navigational equipment; construction staging logistics, etc. The refinery complex would take approximately two years to complete, during which time Olympic planned to employ approximately 2,500 people. Once constructed, the refinery would employ 950 people, consisting of machine operators, engineers, management, and administrative staff. The supertanker terminal would employ 60 people. There would be two fixed-berth supertanker terminals; one for offloading and loading crude oil, and one for loading refined petroleum products. Nine pipelines would form a web of distribution to carry foreign crude oil in and refined products back out. Crude oil would be received for the Durham Point refinery, as well as transshipped elsewhere within the United States. The Durham refinery would process 400,000 barrels of oil per day. About 45 per cent of the refined petroleum products would be distributed back out by pipeline to be transshipped to other New England states and beyond. The largest of the pipelines would be forty-eight inches in diameter, and there would be two of them, capable of handling significantly more than the crude oil needed for one refinery. The underwater pipelines would extend from the supertanker terminals at the Isles of Shoals to the shoreline in Rye at Concord Point, continue underground through Rye to the town of Portsmouth, through Pease Air Force Base, and then underwater across the Great Bay to the refinery at Durham Point.[18]

Unfortunately for Olympic, their final report was not made public until the end of February 1974 at the first of two presentations in Durham. By that time, opposition had the advantage of getting ahead of Olympic's message, winning the public relations battle. Rye residents had already made up their minds. They wanted no part of Olympic's proposal. The underwater blasting off Rye's coastline would devastate the seabed, killing their lobster and fishing economies. The beaches would be blown up as well, destroying their tourist economy. Residents would face a minimum of six months of blasting and pipeline construction activity, stressing the roads, police and fire services and impacting their quality of life. Construction staging areas would disrupt traffic patterns with massive pipeline equipment and materials.

It is important to note, Olympic's plans incorporated many changes over time, particularly as they completed their analyses. One key change was relative to the type of supertanker terminal Olympic would construct. The monobuoy concept was abandoned for two fixed-berth terminals that

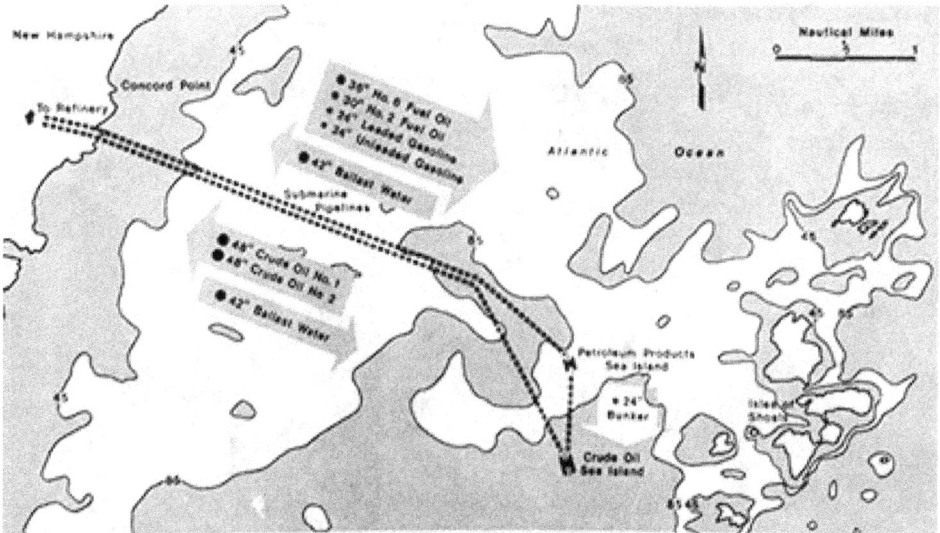

"Proposed offshore marine terminal and diagram of pipelines to the refinery." Olympic Refineries proposal.

Typical two-berth sea island similar to the proposed Olympic marine refineries terminal.

Model of proposed Durham Point refinery. Olympic Refineries proposal.

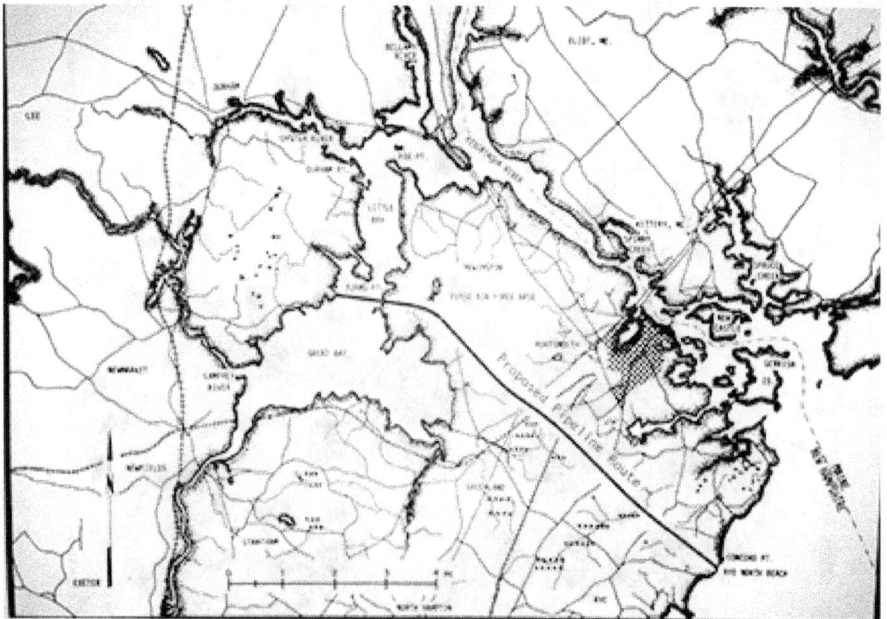

"Map of Southeast New Hampshire and Proposed Pipeline Route."
Olympic Refineries Proposal.

had certain advantages, not the least of which was being able to safely offload crude oil from docking supertankers in the high seas of winter off the coast of New Hampshire. An early conversation with any lobsterman working the Shoals might have spared Olympic some time and money.

The amount of fresh water needed was another major element that changed over the course of time. Olympic's original plan was to use 6,000 gallons per minute (gpm) of fresh water for the refinery. That estimate was reduced to 3,000 gpm, then finally 1,500 gpm, in the face of fierce community opposition and in the course of about five minutes in one public presentation. Olympic's water estimates varied so widely and so quickly, that many began to question the integrity of Olympic's expertise.[19] Of course, Olympic's plan was much larger than one refinery, despite Olympic's dissembling. At the time of Olympic's report, New Hampshire consumed approximately 80,000 barrels of oil per day. The governor hailed the refinery project as necessary for New Hampshire—to control the supply of energy and meet growing demand, bring in good jobs, and fill the state's coffers with millions of dollars in tax revenue.[20] For too long, New Hampshire had been relying on the Gulf and Mid-Atlantic states to supply its energy. The governor's message was also infused with a bit of finger wagging. It was time for New Hampshire to step-up and do its share, particularly now, given the nation's energy crisis.

Of course, simple arithmetic might tell a wider tale. Olympic's refinery complex would have handled 400,000 barrels of oil per day—a figure well beyond meeting New Hampshire's needs. These were initial projections. The two fixed-berth supertanker terminals and nine pipelines would have formed the infrastructure needed to expand that number considerably.[21] Olympic's own consultants estimated "about 45 percent of the refined products would leave the state by the offshore terminal via smaller tankers."[22] That's a lot of oil, and that means a lot of tanker activity. It is hard to imagine what that scale of an operation might look like from the shoreline. As John Kingsbury, director of the Shoals Marine Lab, said in his book *Oil and Water: The New Hampshire Story*:

> Supertankers, oil terminals, and even the ocean environment in which they operate are beyond the usual experience of the average citizen. . . . The exact definition of a supertanker is elusive. . . . These are big boats—the biggest ever built. It is not easy to visualize their immensity. . . . You could almost get three tennis courts to fit lengthwise across the width of a deck of a supertanker. . . . You could place one in Fenway Park (home of the Red Sox) only if you lengthened the distance to the center-field bleachers (Fenway's longest dimension) to three times its present length. . . . On many boats the crew use bicycles to get around.[23]

– 4 –

The Vote...

I n March 1974, several Seacoast towns, including Rye, went to the ballot box to cast their vote on the oil refinery project, among other town matters. The Town of Rye had several questions on the ballot dealing with the refinery project; the legally binding question was Article 4 of the Town Warrant which asked voters if the Isles of Shoals should be included in the Town of Rye's zoning and further that it be zoned as a single residence district. The warrant article was the result of a petition by Charles Tallman who researched the town's zoning map and astutely uncovered an oversight relative to the Isles of Shoals. The town vote was 1,073 Yes (87%) to 194 No (13%) to zone the Isles of Shoals a single residence district, thereby blocking Olympic from using the islands for an offshore deep water terminal or pipeline hub. Neighboring towns of North Hampton and Hampton also decisively rejected the oil refinery project. Together with Rye, the three towns formed a shoreline blockade that would prevent Olympic's pipelines from reaching any inland refinery.[24] In the town of Durham, 90 percent of voters rejected the refinery.

The University of New Hampshire (UNH) also had a role to play in the refinery project. Its campus is located in Durham, very close to the proposed refinery site. At the request of the governor, UNH assembled a core team of experts across many technical disciplines to study the impact of an oil refinery in southeastern New Hampshire. The UNH team was specifically advised by the governor not to address their report to any specific site, and not to offer any conclusions. They also were not provided any funding for their research.

The University of New Hampshire's experts offered comprehensive analysis of the environmental and economic impacts of an oil refinery, marine terminal, and pipelines. Among many findings, the team advised "100 parts per million (.1 ml/1) of crude oil is lethal to the American lobster."[25] The team further warned that "oil spills are inevitable" and that if a massive oil spill occurred, the effects would be felt "as far south as Cape Cod." Particularly disturbing, was their analysis of chronic oil pollution, which experts advised would occur as ships routinely unloaded and reloaded crude and refined products harming the coastal marine and animal ecosystems.[26]

From a timing standpoint, UNH's official report was not released until April 5, 1974. The report's release was held up by negotiations with the governor relative to conclusions made by UNH experts. This, of course, was after the town votes in Rye and Durham.[27] Despite the delay in releasing the official report, it is important to remember university experts were neighbors with opposition group members and were impacted directly by the controversy. The technical experts at UNH worked many long hours and played a critical role in helping seacoast residents understand the potential impacts of an oil refinery.[28]

– 5 –

The Power of Community . . .

WHILE IT IS BEYOND the direct scope of this essay, it is important to put Rye's opposition efforts into context within the broader seacoast region. Three key forces were at work that made an immeasurable contribution to the defeat of Olympic Oil. From a community standpoint, the largest opposition group in the Seacoast region was Save Our Shores (SOS), based in Durham. Nancy Sandberg, a twenty-seven-year-old stay at home mother of a young daughter, with no prior activist experience, was asked to chair the organization. She was thrust into the media spotlight, and became the leading voice of a citizen opposition group that grew to over 7,000 members. Sharon Meeker was the SOS public relations and outreach director, with prior experience battling housing development in New Jersey. Sandberg, Meeker, and countless other dedicated Durham residents, many of whom were women raising children, worked relentlessly to oppose the refinery project. Proximity to the University of New Hampshire afforded SOS access to many resources, including technical experts in a broad cross-section of science and engineering disciplines. SOS's efforts led to the Town of Durham's 9:1 vote against the refinery project.

Nancy Sandberg and her four-year-old daughter Betsy holding the SOS petition of Durham voters opposed to the 400,000 barrels per day Olympic Oil refinery for Durham. The petition stretched from one end of Main Street to the other. Shortly after, State Legislator Dudley Dudley presented the petition to an angry Governor Thomson who promptly told her to get out of his office. In December of 1973 Durham photographer L. Franklin Heald took this photo which is in the book published by the Durham Historic Association entitled **Durham New Hampshire: A History 1900-1985**, *Phoenix Publishing, Canaan, NH, 1985. Courtesy of Durham Historic Association.*

On the legislative front, a first-term state legislator representing Durham, Dudley Dudley, led a fierce effort to protect the town of Durham's right to self-governance. She took the governor head-on and was successful in getting House Bill 18 passed in a special legislative session. Known as the "Home Rule" bill, it codified "the Yankee tradition that every community has the right to determine its own zoning laws."[29]

On the media front, a startup newspaper, *Publick Occurrences*, broke the story and covered it relentlessly, pressuring the governor and Olympic Oil to be transparent with the people. The publisher, Phyllis Bennett, was a seasoned journalist, most recently with the *Baltimore Sun*. She cobbled together an eclectic team of reporters who wrote intelligent, persuasive, and insightful stories that kept the Seacoast communities informed and connected throughout the controversy.

The concentric efforts of these women, and all the activists in their community, helped defeat the entire refinery project as well as reorient the state's position on future coastal zone development.[30]

New Hampshire State Legislator Dudley Dudley. Courtesy of Dudley Dudley.

Phyllis Bennett, publisher of Publick Occurrences *newspaper, which broke the story November 2, 1973. Courtesy of the Bennett family.*

– 6 –

All the King's Horses and All the King's Men . . .

I N THE FALL OF 1973 to the spring of 1974, events moved quickly in New Hampshire, with powerful forces at work, intersecting to develop the oil refinery project. The newly elected governor of New Hampshire was politically ambitious and was backed by the bullhorn of William Loeb, the acerbic publisher of the only statewide newspaper, the *Manchester Union Leader*. Together they tried to curry Republican Party favor, by supporting the president's plan of developing all sources of energy. Together they promised New Hampshire better jobs and millions of dollars in tax revenue.

Yet, all across the country, the economy was weakening, unemployment was rising, and so was inflation. Americans were enervated by war and tired of government deception. Vice President Agnew resigned in October 1973 amid a scandal of political corruption. A month later, President Nixon was on national television assuring Americans he was not a crook—ironic since he was elected on a platform of law and order. Gas prices were rising and energy shortages vexed all Americans. It was part of what author and scholar

Jonathan Schell called *The Time of Illusion*. Americans' lives were filled with distortion and dissonance. This was the cultural, political, and economic context for the Olympic Oil controversy.

New Hampshire was completely unprepared for an oil refinery project, particularly the size and scale proposed by Olympic. The state had "no laws governing the location, regulation, or taxation of oil refineries."[31] The federal jurisdiction was opaque, at best. The Department of Transportation was responsible for the pipelines, but the Environmental Protection Agency had jurisdiction over any pollution from an offshore marine terminal or refinery. No single federal agency could be identified as responsible for granting a permit for the overall refinery project.[32] The state was just not ready to handle massive industrial development. It needed a comprehensive land-use strategy to properly guide the evaluation, siting, development and regulation of heavy industry.[33]

Of course ambiguity invites exploitation when so much money and power is at stake. Governor Thomson and William Loeb led one of the world's richest men, Aristotle Onassis, into believing his oil refinery project could get done. Together they tried to make seacoast residents feel that supporting an oil refinery project was somehow patriotic.

Unfortunately for Olympic, things did not turn out as they had expected. The Olympic refinery project was dead before any application for a permit was even made. The Rye and Durham March votes were a landslide against the refinery project. North Hampton and Hampton shored up the battle line. Concerned Citizens of Rye, Save Our Shores, and other citizen opposition groups across the Seacoast mobilized and unified their efforts to preserve their livelihood and quality of life before Olympic had even completed their final public presentation. The Dudley "Home Rule" bill united political adversaries in the state legislature under an umbrella of tradition. The Remick sisters and Rye's determined opposition closed off a critical pipeline route. Rye citizens rezoned the New Hampshire Isles of Shoals out of Olympic's reach. Star Island Corporation sent Olympic back to the drawing board to try and come up with another site for critical navigational equipment. There was only so much maneuvering Olympic could do to build a marine terminal in water deep enough to safely handle supertankers, yet staying within the three-mile zone controlled by the governor.

After the town of Durham rejected an oil refinery, other towns stepped into the debate. Newmarket and Rochester were the only main alternatives under consideration. Their town residents voted in favor of a refinery in late April— jobs and tax revenue being the allure there. None of these overtures ever got traction. Given the support these towns

Isles of Shoals
may get zoned

The Isles of Shoals, currently a target of those interests anxious to put an oil tanker port on this coastline, is now the subject of a zoning puzzle in Rye.

The Town of Rye owns Star, Lunging and White Islands but does not include them on its zoning maps.

This week the Rye Planning Board was asked to place the islands on the map as a single residence district.

The request came in the form of a petition with 50 Rye signatures. Only 25 signatures are needed to have a zoning ordinance presented or an existing code considered for amendment by public vote.

Charles V. Tallman, a former chairman of the Rye Planning Board, posed the original question to the planning board recently and learned that the islands were not on the map.

"Isles of Shoals may get zoned." Publick Occurrences, *January 11, 1974.*

Boston Globe

March 6, 1974

Rye zones an island out of Onassis reach

RYE, N.H. — Residents here last night voted overwhelmingly to zone Lunging Island, optioned by Onassis's Olympic Refineries for a tanker terminal, for residential use.

North Hampton residents, also voting in a town meeting, rejected a proposal for a refinery in the New Hampshire seacoast region by a unanimous voice vote with some 500 persons present.

By a vote of 1073-194, Rye voters approved a legally binding measure to restrict the island, one of the Isles of Shoals, for residential use.

Aristotle Onassis's company had proposed the island to be used for tanker operations as part of a proposed $600 million oil refinery complex involving Durham, Portsmouth and Rye.

On three, non-binding questions, voters in Rye also overwhelmingly opposed locating an oil refinery, offshore oil facilities and pipelines in the seacoast region.

Olympic Refineries presently plans to run a network of pipelines under the seacoast town of Rye which would connect the proposed 400,000 barrel-a-day Durham Point refinery to tanker terminals planned eight miles off the New Hampshire coast.

Rye has control of the three New Hampshire Isles of Shoals, including Lunging Island, originally scheduled to be used by Olympic Refineries for the tanker operation.

Now, the oil firm is suggesting that two tanker terminals be located west of the Isles of Shoals.

Boston Globe

March 6, 1974

Durham still Olympic's refinery site

By David Rogers
Globe Staff

Olympic Refineries yesterday reaffirmed its desire to build in New Hampshire and denied published reports that it may drop plans for property in Durham, N.H., where the company has met opposition, in favor of a site in Melville, R.I.

Durham residents are scheduled to vote tonight on a resolution on the refinery issue, but Constantine Gratsos, an assistant to Olympic's owner, Aristotle Onassis, said yesterday that the company intends to continue with its plans there whatever the outcome of tonight's town meeting.

United Press International yesterday reported that Gratsos had indicated in an interview Monday that Olympic might move to a tank farm in Melville, R.I., near the old Newport Naval Base, but the executive said his remarks had been misinterpreted.

"The conversation was merely informative," said Gratsos. "They asked me if it (Melville) is a good site and I said 'yes and we would be interested if we decided to build a second refinery in New England."

UPI also reported that Gratsos or another Olympic spokesman would visit Rhode Island next week to meet with Gov. Phillip Noel on the refinery question, but Gratsos said no such plans have been made.

Gratsos indicated he may be in New Hampshire tomorrow after the Durham town meeting. The visit would coincide with an expected vote in the state's Legislature on a measure which would allow a state site evaluation committee to overrule local zoning powers.

The Durham town meeting resolution asks only how the residents feel about refinery construction in the town. It is too ambiguous to be legally binding and is considered as essentially a test of public opinion.

Another referendum would be needed for approval of zoning changes which Olympic needs to build a refinery.

"Rye zones an island out of Onassis' reach." Boston Globe, *March 6, 1974.*

showed for a refinery, it is evident Olympic couldn't quite make the rest of the critical pieces work—specifically, the supertanker terminal and pipeline complex. No crude oil, no refinery.

Olympic, and even the governor, with his southern drawl exposing his Georgian roots, somehow misunderstood the steadfast Yankee ethos of self-determination and self-governance. Onassis and his team would learn that money doesn't always triumph over tradition. It's interesting to note, Olympic paid a consultant to explore the history of Durham and the Great Bay estuary. It's unfortunate they didn't explore Rye's 350-year history of fierce independence. Perhaps a great deal of time and money could have been spared. If nothing else, Olympic came to understand the meaning of General John Stark's words that form the state's motto: "Live Free or Die."

In the final analysis, Olympic Oil was defeated by the fierce determination of separate, yet interrelated and interdependent, Seacoast communities. Like nature itself, it was an ecosystem—each town played a vital part, and each town was made stronger by unity with the other. Olympic ultimately returned to New York, feeling it had been badly used by New Hampshire. For Aristotle Onassis, it all ended rather like a Greek tragedy—he died on March 15, 1975, at the age of 69.

Epilogue

THE OLYMPIC OIL controversy in 1973 intersected with a transformative time in America's cultural and environmental history. Beginning in the 1960s, Americans began to question the wisdom of modern science and its perceived advancements in controlling nature. Seminal works such as Aldo Leopold's *A Sand County Almanac* and Rachel Carson's *Silent Spring* reshaped the way many Americans perceived land use and the interrelationship of all living things. In 1969, the Santa Barbara oil spill occurred. Americans were shocked to see 200,000 gallons of crude oil cover 800 square miles, blackening 35 miles of coastline.[34] Shortly thereafter, on January 1, 1970, the National Environmental Policy Act was signed into law "recognizing the profound impact of man's activity on the interrelations of all components of the natural environment, particularly the profound influences of population growth, high-density urbanization, industrial expansion, resource exploitation, and new and expanding technological advances...to create and maintain conditions under which man and nature can exist in productive harmony."[35]

On April 22, 1970, the first Earth Day was celebrated by millions of Americans concerned about oil spills, industrial contamination, pesticides, air pollution and water contamination. This movement gave rise to the creation of the Environmental Protection Agency, designed to protect people from environmental hazards. The Clean Air Act and Clean Water Act are just two examples of federal environmental legislation that came about in the early 1970s.

Americans increasingly viewed natural resources as part of a larger ecosystem and in need of protection from exploitation. This environmental history sets the subtext for opposition to Olympic Oil and its efforts to exploit the Seacoast region of New Hampshire for profit.

In 1973, geopolitics wreaked havoc on the nation. Disruptions in petroleum sparked misery for New Hampshire. For the Seacoast, misery was just the beginning. The Seacoast was battling an oil refinery complex *and* nuclear power plant courtesy of the Thomson administration. To the eternal regret of opponents, the Seabrook Station was ultimately constructed. Some argue the controversy over the oil refinery complex drained resources that were stretched too far to effectively win the battle against the nuclear power plant. The lead developer of the Seabrook Station, Public Service Company of New Hampshire (PSNH), faced protracted legal challenges and acts of civil disobedience from activists such as the Clamshell Alliance, cofounded by the late Guy Chichester of Rye, and the Seacoast Anti-Pollution League (SAPL). Ultimately, the nuclear station was constructed, but PSNH went bankrupt and the project was limited to one reactor (two were proposed).

Today, the state of New Hampshire relies on the Seabrook nuclear power plant to generate about fifty percent of its electricity. It is a legacy of the Thomson years and remains a source of controversy and debate. There is little to suggest that nuclear won't continue to play a major role in electricity generation for the state well into the future. Indeed, the majority owner, NextEra Energy, is seeking extension of Seabrook Station's operating license from the Nuclear Regulatory Commission to the year 2050.

The challenges of this era may bear a different name, but they are rooted in the same inexorable hunt for energy and pursuit of profit from the land. The Northern Pass and SEA-3 are just two examples of energy projects being heavily debated today. The fact is New Hampshire does not produce petroleum. It does not produce natural gas. And it does not refine petroleum. Yet, it has a population of approximately 1.3 million people, half of which rely on petroleum as their primary heating fuel, according to the U.S. Energy Information Administration (EIA). In large part, New Hampshire continues to rely on the mid-Atlantic and Gulf states, as well as imports from Canada, to meet its petroleum needs. Portsmouth still does the heavy lifting for the state, receiving and storing refined petroleum products for further distribution by truck and rail.

History teaches us change is constant and the past has a way of repeating itself. New Hampshire is striving to reduce its dependence on petroleum products. In July 2013, the current governor, Maggie Hassan, signed into law Senate Bill 191 which directed the Office of Energy and Planning to create a ten-year energy strategy for New Hampshire. The Office of Energy and Planning released its final report in September 2014 focusing on "energy efficiency, fuel diversity and transportation."

The report is available at https://www.nh.gov/oep/energy/programs/documents/energy-strategy.pdf. The state has a "Renewable Portfolio Standard" requiring that approximately twenty-five percent of its electricity come from renewable sources by the year 2025. In 2015, the EIA reported "most new renewable projects under development in New Hampshire are powered by wind or biomass." Indeed, the "first modern wind farm opened in 2008, and more than half a dozen additional projects are operating or are in development." Residents of the Seacoast will note the EIA's observation that to date "there have been no proposals for turbines off New Hampshire's 18-mile Atlantic coastline."[36]

Appendix A

Community Reflection . . .

by Peter Horne and Friends

M

Y RECOLLECTIONS, and those of my Rye friends, Betsy Bischoff, Mike Flanigan, and Tom and Marion Barron, are included here. There were other friends and neighbors engaged in the opposition to Olympic Oil and we thank them for their friendship and efforts. We are offering this recollection with "one voice."

When we look back to 1973-74, forty plus years ago, we are reminded of how we needed "all hands on deck" to stop the terrible attempt to turn the Shoals and the Seacoast into a big oil port. We knew if Olympic Oil succeeded in their efforts, the town of Rye as we knew it would be destroyed. We hope this reflection will give the reader an "on the ground feel" of the dynamics in play at the time.

The recollections we share here are from events that occurred during the period from October 1973 to September 1974, when Olympic Oil finally went away for good.

In October 1973, Betsy Bischoff received a telephone call from Sharon Fish, a Rye resident and friend. Sharon told Betsy there was a meeting in Durham at the Durham Community Church to discuss a development in the area—real estate agents were approaching homeowners on Durham Point and offering options to buy their land. Rumors were flying that these out-of-town realtors were representing different people, to create horse farms, game reserves and protected open space. All of which turned out to be lies! Sharon and Betsy attended the meeting at the Durham church. Sharon then suggested to Betsy that she hold a meeting at her home in Rye and invite neighbors to discuss the developments. Betsy's home was under renovation at the time, so everyone sat down on piles of lumber that were drying out in her living room. It was at this meeting the decision was made to form an ad hoc committee to dig for more information. Jessie Herlihy hosted the next meeting at her home, to keep the ball rolling and focus on what to do next.

Betsy Bischoff and Mike Flanigan then joined a meeting in Durham Point hosted by Nancy and Mel Sandberg. The Sandbergs lived on Durham Point and had been approached by one of the realtors trying to option their land. It wasn't long after this meeting that Save our Shores (SOS), a Durham-based grassroots opposition group, was formed. In Rye, folks gathered to discuss the idea of forming our own organization. Peter Horne, Betsy Bischoff, Judy Paterson and Mike Flanigan traveled to Durham to meet with SOS members, including Sharon Meeker, who was in charge of community outreach. Sharon Meeker suggested that Rye join SOS, rather than form its own organization. Our sense was a grassroots group of Rye citizens could be more effective in fighting off the proposed oil terminal and pipelines, while SOS fought the oil refinery proposed for Durham Point. We had a lot of common ground with Durham and closely collaborated on issues such as hearing from the experts at UNH and "Home Rule" which was being introduced by Durham's state legislator, Dudley Dudley.

The governor and Olympic Oil left the Seacoast with more questions than answers. We will be forever grateful to the late Phyllis Bennett, publisher of the start-up community newspaper Publick Occurrences. Phyllis and her team of reporters covered every angle of the story to inform residents and shine a light over Olympic's plans for all to see.

Rye residents were hungry for information about the project and of course devoured every copy of *Publick Occurrences* that was mailed to their homes. Everyone was on edge. Our first big effort was to organize an information session to start getting answers to our questions and inform the public of the pros and cons of Olympic's plans. The event was to be held January 23, 1974 at the Rye Junior High School gymnasium. Peter volunteered to be moderator. The meeting included representatives of Olympic, a panel of experts from UNH, including John Kingsbury from Cornell who ran the Isles of Shoals research station, and a representative from the federal Environmental Protection Agency (EPA), who specialized in offshore oil terminals. The meeting was attended by a great mix of about a thousand Rye citizens to learn more about Olympic's plans. We all hoped this meeting would shed some light on this huge project.

And then an extraordinary coincidence happened—to this day it rings in our memories! An airline pilot named Kit Baker lived in Rye. He told some friends that he had a layover coming up in New York and when he had some free time he would drop into Olympic's office to see if he could find out more information about their plans. Kit Baker did just that. The people in Olympic's office told Kit that they were too busy to meet with him, but offered him a film to view. The film described how pipelines were built in the Middle East. Upon conclusion of the film, Kit told his host that it was a good film and they should bring it to Rye. Sure enough, for whatever reason, the Olympic representatives brought the film with them to the information session at the junior high school. It brought the house down! The narrator spoke in a British accent and at one point said, "Occasionally we run into stubborn rock!" A huge explosion followed the narrator's comments, as Rye residents watched rock and sand blow high into the air! The next scene in the film was of a large construction project in which the workers were assembling four-foot diameter pipes on the beach. These memorable scenes were etched in everyone's minds, including our lobstermen who worked out of Rye Harbor. There is no doubt that blast galvanized the opposition in Rye. Although there was a lot of work ahead of us, the message was clear: We had to stop this project!

Shortly after the January information meeting, we met as a group to complete our formation of Concerned Citizens of Rye (CCR). We went to work, getting the word out to our fellow citizens about what CCR stood for, and what our policy and mission would be. We spent considerable time discussing the tasks before us and developed these core values:

- A shared regard for the history of Rye
- A commitment to the protection of land and sea and our clean beaches
- To protect the livelihood of our lobster fishermen
- Our vision of Rye as a residential and recreational community
- A community with a sense of time and place for its citizens
- A positive relationship with our sister communities that make up the Seacoast
- And an overall determination to fight hard as a town for the values we hold dear

We then formulated our statement of policy and intentions and posted them in our newsletter, *The Coast Watcher*. Tom Barron printed the newsletter in his kitchen out of his house on Washington Road, along with Walter (Snick)

Wilson. Guy Chichester, Judy Paterson, Walter and Peter shared in producing much of the written commentary. Mel Low and his daughter were among the volunteers who delivered the newsletter to every mailbox in town. It was one heck of an effort!

When we formed CCR we decided to use a consensus-building approach to establish priorities for action rather than a top-down structure. We had a lot of expertise within our organization and members volunteered to take on tasks they could successfully complete. Peter volunteered to be Chair and Judy Paterson volunteered to be Co-Chair to head up our efforts on the "Home Rule" issue in collaboration with SOS and Dudley Dudley's legislation. Some of us recall being called "crazy" to be part of a group going against Aristotle Onassis, Governor Thomson and *The Manchester Union Leader*. In time, many seacoast residents who supported the refinery project gradually changed their minds. *The Coast Watcher* and letters to newspapers from experts in the field helped educate the public with facts regarding blasting, lack of fresh water required, etc.

CCR also recognized the importance of support from friends in neighboring seacoast towns. Tom and Marion Barron were active in outreach activities. The Barrons contacted Ed and Mary Loughlin who were motel owners in Hampton Beach and an important part of CCR's support in Hampton. The Barrons also contacted John Dineen and his son about having a benefit in the spring at the Hampton Beach Casino. John Dineen invited all the members of CCR and those opposed to Olympic to attend an evening with the Count Basie Band at the casino free of charge as a thank you for our work on behalf of the Seacoast. It turned out to be a grand night and well attended.

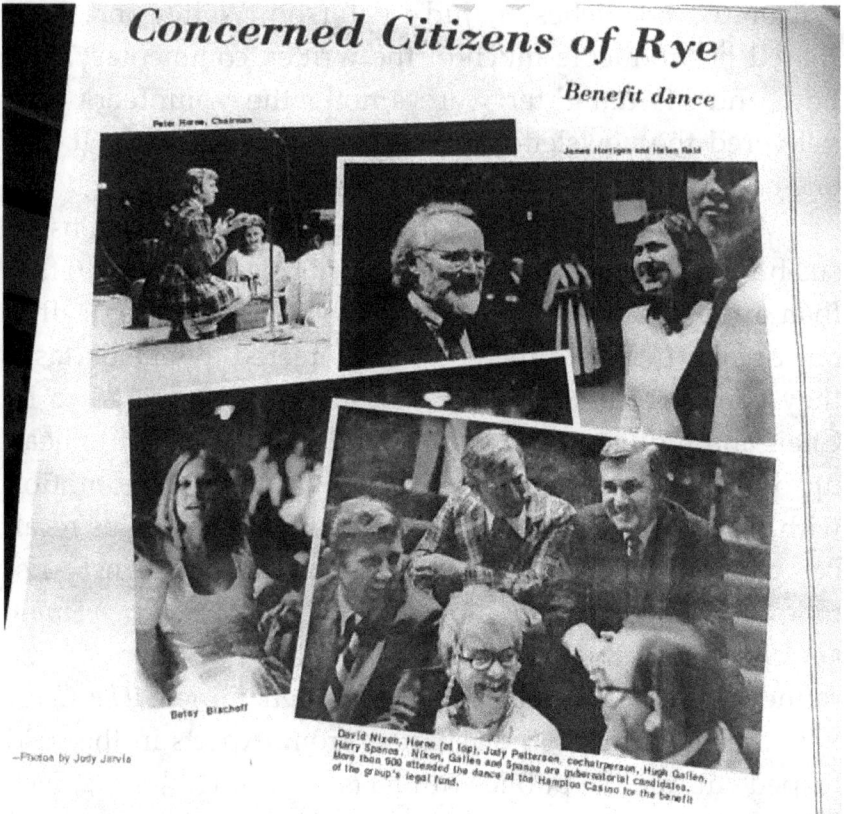

Concerned Citizens of Rye

Benefit dance

Pete Horne, Chairman

James Hartigan and Helen Reid

Betsy Bischoff

—Photos by Judy Jarvis

David Nixon, Horne (at top), Judy Patterson, cochairperson, Hugh Gallen, Harry Spanos, Nixon, Gallen and Spanos are gubernatorial candidates. More than 900 attended the dance at the Hampton Casino for the benefit of the group's legal fund.

As time passed we gained momentum, but more complicated issues had to be addressed. We witnessed the wonderful intellectual savvy and commitment of Louise and Charles Tallman and others who were members of different municipal boards and volunteers. There was unity between us, and a whole lot of collaboration by Rye volunteers who served on various municipal boards in our town.

We also discovered the precious values held by the Remick sisters, which to this day continue to be a lesson for us all given the constant pressure on our environment by industrial projects.

Rye was doing its part to defeat the refinery project, but we did not know if it would be enough to eventually prevail over Olympic and the governor's plans. The outcome was particularly uncertain after the town of Durham rejected the refinery, but the towns of Newmarket and Rochester stepped in to bring Olympic into their communities. We knew we needed to maintain our vigilance until Olympic Oil went away for good!

Our emotions ran deep over the many months the threat was before us and we gathered with gusto when we had the opportunity at Jarvis's Market Square Pub in Portsmouth where the Shaw Brothers sang. Doug Gray would sing his favorite Irish tunes and the Shaw Brothers would lead us in a sing-along of our battle hymn "Ode To Olympic" as the night ended. Those evenings were a great source of entertainment and camaraderie. The Shaw Brothers contributed a lot to our emotional spirit and fueled our determination to get it right for Rye.

Shaw Brothers. Courtesy of the University of New Hampshire. Photo by Lisa Nugent.

Ultimately, we defeated Olympic Oil and the governor. It took "all hands on deck" to make it happen. Now, many years later, we have reflected on this event. There were many, many, good folks involved in the effort. Too many of our friends are sadly no longer with us. With this publication, we honor all the people who helped preserve Rye, our seacoast, and our way of life. We also offer this reflection as a "Lessons Learned" to the next generation, so they may continue to be good stewards of our beautiful town and seacoast. It is in this spirit, we offer our "Lessons Learned":

- Determine what your organization stands for before you engage in opposition to an issue
- Do not be secretive
- Be active in learning about complicated issues
- Try to get to the truth of the matter
- Make full use of science and expertise available to you
- Keep emotion in check, and listen to others
- Be well organized and share the tasks
- See what expertise is in your community
- Take full advantage of communication systems in your community
- Do not be intimidated
- Keep your sense of humor and add some music and fun to your efforts

APPENDIX B

On a Lighter Note...

Ed Valena produced a musical about the Olympic controversy called "Oiley-Vey" that was a comic triumph in Durham. The town of Durham also has a 1500-pound granite bench in "Wagon Hill," overlooking the Great Bay, memorializing this event. It is inscribed, "March 1974 Durham says NO to Olympic Oil Refinery." Rye does not have a granite bench, but if you listen closely, you just might hear its battle hymn, *Ode to Olympic*, sailing in with the tide from the Isles of Shoals.

"Ode to Olympic"

One night a man named Booras
Came driving into Rye
He laid his plans before us
And tried to make us buy
An oil-refinery city
Great pipelines it would be
With engineers and Arab ideas
For everyone to see

(CHORUS)
For it's Rye, Rye, a salty old town,
Great Beaches, the ocean and
Shoals
For it's Rye, Rye, we won't let you down
To Hell with Olympic and oil

A thousand townsmen gathered
And listened to his plans
And it wasn't very long
Before they could hardly stand
The thought of giant tankers
A-foundering on the Shoals
With Bunker C for you and me
A-flowing from their holds.

The Isles of Shoals and fishing
Would certainly be gone
And it is no fun to sail
On Olympic's settling pond
The beaches would be blackened
And Rye would only be
A very very brief recollection
In our children's memory

Now listen, my fellow citizens,
And listen very well
The future of our seacoast
Should not be up to Mel
So let's tuck it to Olympic
And "El Supremo" too*
And from the citizens of Rye
We say to hell with you!

*Note: The author, Peter Horne, was chairman of Concerned Citizens of Rye. He submitted this to *Publick Occurrences*, which edited the last verse. The original unedited text is presented here. Reprinted with permission.

APPENDIX C

Photographs

Tanker collision. Courtesy of Rye Town Museum. Committee for Regional Oil Planning, Arthur E. Martin, undated. "It could happen at the Isles of Shoals."

Aerial view of the Isles of Shoals. Photo courtesy of Rye Town Museum.

Fishermen of Rye. Photo courtesy of Rye Town Museum.

Tom and Marion Barron. Courtesy of Tom Barron.

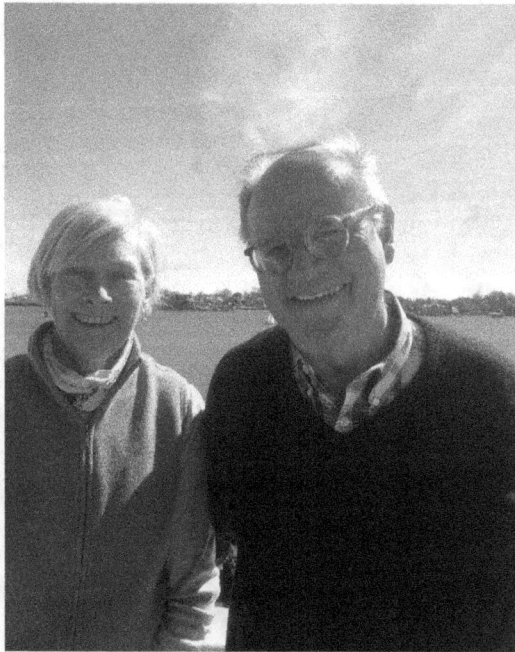

Bill and Betsy Bischoff. Courtesy of Betsy Bischoff.

Guy Chichester. Courtesy of Rye Town Museum. Madeline and Guy Chichester. Courtesy of Madeline Chichester.

Mike and Sally Flanigan. Courtesy of Sally Flanigan.

Jessie Herlihy. Courtesy of Alex Herlihy.

Peter and Holly Horne. Courtesy of Peter Horne.

Bill and Priscilla Jenness. Courtesy of Priscilla Jenness.

Mel and Jean Low. Courtesy of Jean Low.

Dave and Judy Paterson. Courtesy of Sharon Paterson McKinney.

"Sister Soldiers." Frances Tucker and Bernice Remick. Publick Occurrences, *March 1, 1974.*

Charles and Louise Tallman. Courtesy of Rye Town Museum.

The Governor and Onassis

'I am no Greek bearing gifts'

The governor and Onassis. Photo credit: Judy Jarvis. Publick Occurrences, *December 21, 1973.*

Photos of Jay Smith

These are three of a number of photographs by Jay Smith which dramatized the struggle between Olympic Refineries and Durham, the town which it chose for an oil refinery.

When Governor Thomson announced the refinery plan and gave his endorsement to it, Smith caught the governor, like an aide-de-camp, his arm on the Governor's chair in which Peter Booras was sitting, a symbolic reminder of how major industrial concerns can seat themselves in political power.

At the Durham town meeting, Alden Winn, chairman of the selectmen, set the defensive mood — and took the role of hero — by charging Olympic with deception.

Governor Thomson and Olympic Oil representatives. Photo credit: Jay Smith. Publick Occurrences, *August 23, 1974.*

"Papanicolaou: Olympic's Ambassador" Publick Occurrences, *March 8, 1974. (Pictured are: Peter Booras, Governor Thomson, and Nicholas Papanicolaou.)*

William Loeb III, publisher of the Manchester Union Leader, *1973. Photo Credit: Spencer Grant. Boston Public Library. Reprinted with Permission.*

Aristotle Onassis. Photo credit: Nationaal Archief Collection.

...now to make the plant blend in with its surroundings...

Bob Nilson cartoon. Publick Occurrences, 1974.

Bob Nilson cartoon. **Publick Occurrences, 1974.**

The editorial cartoons of Nilson

A good political cartoonist is genuinely rare — a small percentage of the number of successful ones in the country.

Only the very best rise above the task of illustrating an editorial and make independent, forceful statements of their own.

And only the very best, draw so well that they distill intricate ideas clearly without the use of clichés of speech and picture — dull aphorisms, shopworn Republican elephants and Democratic donkeys, and little labels stuck about saying "Liberty" or "The Economy."

Finally, only a handful of the very best have a real sense of humor and are able to present their ideas in that most effective fashion.

Herblock is such a cartoonist, so is Bill Mauldin, and so is Bob Nilson of Publick Occurrences.

The most successful of these dealt with the proposal by Aristotle Onassis to build a monster oil refinery in Durham.

Durham did not exactly look upon the plan as a gift.

Some of Nilson's best ideas came back recycled when an original event took a new turn.

Two birds in a nest made the first statement on the inappropriateness of the proposal. And the same two birds returned to signal Durham's final answer.

"IT'S OUR NEST ALL RIGHT. BUT IS IT OUR EGG?"

When it was learned that the refinery was being proposed by a Greek firm, Nilson called on the Trojan horse to depict the deceptive ways in which the proposal was being put into action.

When a later story detailed how oil-rich Arabs were secretly bankrolling new refinery projects, Nilson returned to the Trojan horse again.

Bob Nilson cartoons. Publick Occurrences, *August 23, 1974.*

Numb. — 12

PUBLICK
OCCURRENCE

November 2, 1973   *The New Hampshire Seacoast Community*   ⁻ents

Newmarket battles Oyster River	Theatre-By-The-Sea's opener		The local drive to get Nixon	Lega over

Options on Durham Point
for over 1000 acres

Eleven Durham Point landowners acknowledged this week that they have granted a total of over 1,000 acres in land options to a Nashua real estate dealer, George Pappademas.

In a telephone survey by *Publick Occurrence*, it was found that Pappademas is making considerable headway in his effort to sew up options on the Point. However, a frequently cited figure of 2,000 acres in options could not be confirmed.

Pappademas is concentrating his activity on the interior of Durham Point, and showing only a casual interest in shore frontage. According to landowners, he is attempting to put together a pattern of purchases. While he seems unconcerned whether the land is swampy, ledgy, open or wooded, he is trying to purchase parcels of land which fit together into a large block.

Pappademas, who, according to Lane Stoudt of the Nashua Board of Realtors, has been a well-known businessman in Nashua for years—and who has held a real estate license for something over a year—has been active on Durham Point for over 2 months.

Although Pappademas declined to discuss his plans this week, he has indicated to some Durham Point residents that he will make a public statement in the near future.

None of the many people interviewed were able to shed light on the use—if any—to which the land will be put. Nor was anyone able to say with certainty how many options will actually be exercised. Pappademas himself said in an October 3 interview that he had obtained seven options. He said the land would be used for his own investment, for a game preserve for friends in Keene and possibly a family compound.

Views expressed this week on the matter ranged from indignation to concern to wonderment to amusement. Alden Winn, chairman of the Durham Board of Selectmen, is watching the Point closely.

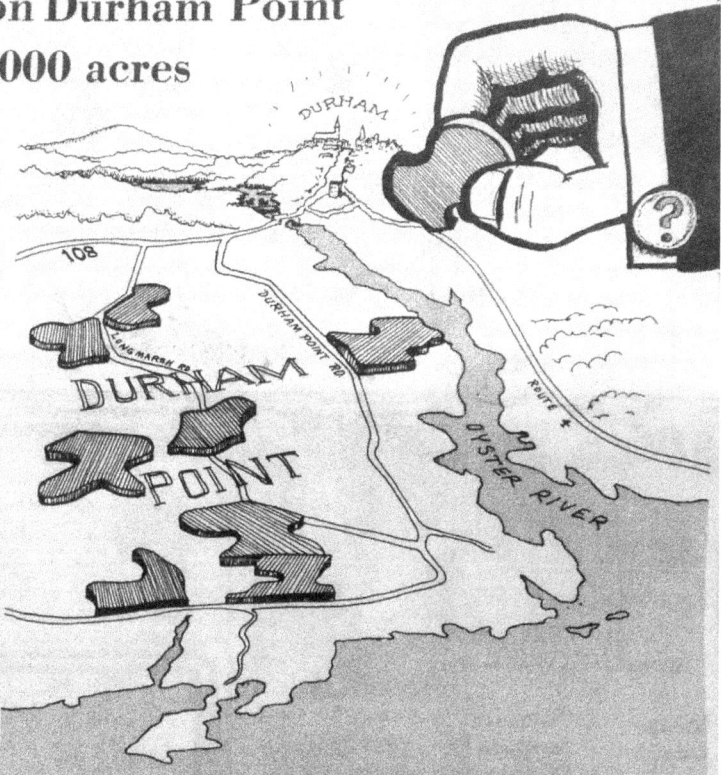

BOB NILSON

Publick Occurrences *broke the story with this headline "Options on Durham Point for over 1000 acres." November 2, 1973.*

LAWRENCE EAGLE-TRIBUNE, Lawrence, Mass.—Saturday, November 17, 1973

Rye, N.H., to get pipeline?

RYE, N.H. (UPI) — Rye residents said Friday they suspect land options being taken in their seacoast township are for a pipeline to feed an oil refinery Gov. Meldrim Thomson indicates will be built in New Hampshire.

Thomson told a governors' meeting in Boston Thursday and repeated to New Hampshire city officials Friday that he expected to announce within two weeks that an oil company would be locating in New Hampshire. He said it would be bigger than the refinery Gibbs Oil has said it will build in Sanford, Maine, 15 miles from the New Hampshire line.

Mrs. Louise Tallman of the Rye Conservation Commission said Friday night a number of residents of Rye, population 4,113, have been approached to grant options on their land.

"Landowners are very considerably upset to be approached, with pressure, and not to be told why. All that we know is the pattern of those who've been approached. I've pieced it together and it runs at right angles to the ocean," Mrs. Tallman said.

Real estate agents said privately they had heard a consortium including Nashua realtor George Pappademas, who lives in Hampton Beach, was taking options on land on Durham Point, just up the Piscataqua River from Portsmouth.

Rye is just south of Portsmouth. It is about 12 miles in a direct line from Rye Harbor to Durham Point.

Thomson has pushed for an oil refinery in New Hampshire since he took office in January. He has mentioned the possibility of a refinery somewhere inland, fed by a pipeline from a floating harbor somewhere off New Hampshire's 18-mile coastline.

The Sprague Oil Co. has nearly completed an oil separation plant in Newington, just upstream from Portsmouth, and has talked of building a small refinery next to it at a future date.

Sprague, Gulf, Gibbs, Exxon and Northeast Petroleum all said Friday they were not the company to which Thomson referred this week.

"Rye, N.H. to get pipeline?" Lawrence Eagle-Tribune, *November 17, 1973.*

Numb. — 15

PUBLICK
OCCURRENCE

November 23, 1973　　　*The New Hampshire Seacoast Community*　　　Cents

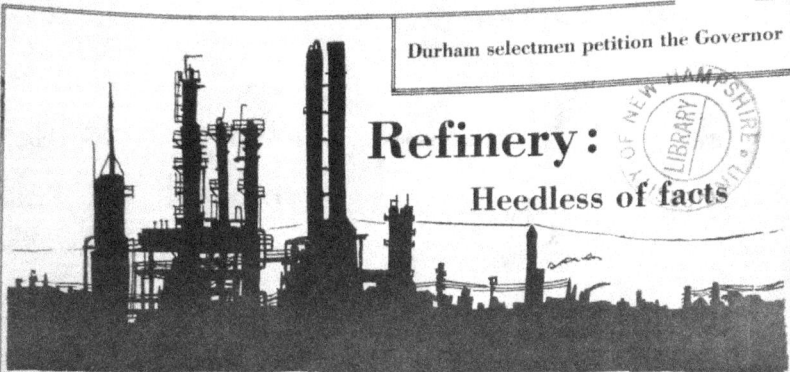

Durham selectmen petition the Governor

Refinery:
Heedless of facts

A checklist of things to find out

Louise Durfee, a Providence, R. I., lawyer, led a successful fight against construction of a refinery in Tiverton, R. I. For *Publick Occurrences* she listed a number of things which she thinks should be considered in planning for a refinery:

Effect on water quality in the area;
Effect on the air (particulate matter, gases and odors);
Existing regulations and means of enforcement;
Penalties for non-compliance;
Effect on the fishing industry;
Effect on the tourist industry;
Impact on local communities;
Traffic problems caused by trucks;
Likelihood of satellite industries (paint, solvent, plastics);
Possible negative effect on the interest in clean industries in the area.

Durham asks
What's going on, Governor?

Durham selectmen planned to send a letter Friday to Governor Thomson asking that he explain what, if any, connection exists between recent large scale land option purchases and a proposed oil refinery for the state.

Alden Winn, Chairman of the selectmen, drafted the letter. He paraphrased its basic message in an interview:

"We have the understanding from George Pappademas, real-estate man who has been picking up the options, that he does not intend any use for the land which would violate our zoning regulations. In light of this, is there any connection between his purchases on Durham Point and your plan to locate an oil refinery in the state?"

He declined to show *Publick Occurrences* the actual text of the letter before it could be received by the governor.

The Durham zoning ordinance clearly prohibits any kind of industrial use in that part of town, Winn said, and variances can be granted only after duly warned public hearings and only if the landowner can prove hardship.

In a statement released Wednesday, Winn said that "since town governments are the creation of the state government there have been speculations that a land zoning ordinance can in some way be overturned by state government action."

He said he has been advised that except in extraordinary circumstances a local zoning ordinance can only be made inoperative by an act of the legislature.

At the regular selectmen's meeting Monday night, Selectman Owen Durgin suggested the town seek outside legal advice "in the unlikely event that the state intends to countervene our zoning ordinances and take our land..."

It was later announced the town has retained the services of Joseph Millimet, senior partner of the firm of Devine, Millimet, McDonough, Stahl & Branch of Manchester, to represent its interests in the land-purchase matter. Millimet also represents the University of New Hampshire, Winn said, and some of the optioned land abuts university property, so there is a combined interest.

At Monday's meeting, the selectmen were most anxious to state publicly that, while they were aware of all the speculation about a refinery, they don't have any more information about it than anyone else who reads a paper.

"A lot of people have called and suggested that we know more about this thing than we're telling," Winn said. But he insisted that it was not so.

Malcolm Chase, who is a landowner on Durham Point, also maintained that he was as much in the dark as everyone else.

Lawrence O'Connell said that the people should know that the town has adequate zoning restrictions and that the selectmen fully intend to enforce them.

Governor Thomson is planning an oil refinery for New Hampshire but he has made no serious impact study of its consequences.

His sole action has been the hasty establishment of an ad hoc committee with no money and few resources to do an in-depth study.

Regional planners say they get no help from the State or from industry and are hampered equally by a lack of resources.

There is a critical lack of information about the environmental effects of refineries. In many areas of concern, the fact is that no studies have been made and no one can speak knowledgeably.

And yet, the announcement of a refinery has been planned for this week.

Regional planners say
they get no help from
the State or industry

Shouldn't the State make an in-depth study of what a major oil refinery would do?

"You mean one of those studies that would take months and cost the taxpayers thousands of dollars?" That was the reaction of Jay McDuffey, the Governor's press secretary.

Instead, the Governor has requested that the University of New Hampshire undertake a 90-day, unfunded impact study in cooperation with the State Department of Resources and Economic Development (DRED).

McDuffey said that this UNH study would be adequate. He called the study group "an important and integral part of the effort to bring sufficient oil supplies to New Hampshire."

However, Robert Faiman, vice provost for research, said of the study group, "We don't have the resources to do an in-depth study."

In addition, he said that Fred Goode, the assistant appointed by the Governor to meet with the group, had no research to pass on to the group.

In effect, the group is starting from scratch.

Thomas Bonner, UNH president, has overall charge of the study group which includes Faiman, William Henry, chairman of the Resources Development Center; David Olson, director of the Institute of Natural and Environmental Resources; Godfrey Savage, director of the Engineering Design and Analysis Laboratory; and Donald Moore, director of the Center for Industrial and Institutional Development.

The UNH group will concern itself primarily with environmental questions, and DRED will investigate economic factors.

Faiman pointed out that UNH has been called upon to "provide advice and counsel" to the Governor's office.

"The University is in no way an interested party in this matter," Faiman said. "We are a state-sponsored institution with the scientific expertise to provide useful information to the Governor's office."

He emphasized that the group had not been asked to select a site.

According to William Henry, the study will amount to two-man-years of labor. (Two man-years equals the amount of work which one man could accomplish in two years.)

Henry said that 10 to 12 experts from both UNH and DRED will be involved. Henry sees the lack of time as a limitation on the researchers.

New Hampshire regional planners have been attempting to prepare for the eventuality of an oil refinery, but have been unable to obtain the necessary information.

Michael Koutkas, director of the Strafford Regional Planning Commission, said that his office invited officials of Gibbs Oil Company to discuss the impact of a refinery but that to date Gibbs has not cooperated. (Gibbs is building a refinery in Sanford, Maine.)

Otis Perry, assistant director of the Southeastern Regional Planning Commission, said that few refineries have been built in this country in the last few years. Consequently, there have been few opportunities to make model impact studies.

"How do you assess the impact of a refinery without have data? he asked.

Turn to REFINERY, page 3

"Refinery: Heedless of Facts." "What's going on Governor?" Publick Occurrences, *November 23, 1973.*

The Boston Globe

WEDNESDAY MORNING, NOVEMBER 28, 1973

To meet 25% of N.E. needs

Onassis planning $600m refinery in New Hampshire

Onassis firm hopes to build N.H. refinery

★ FUEL
Continued from Page 1

By R. S. Kindelberger
Globe Staff

An oil company owned by Aristotle Onassis plans a $600 million refinery at Durham Point, N.H., Gov. Meldrim Thomson announced yesterday.

At a news conference in Concord, Thomson and project coordinator Peter J. Booras said the refinery would provide a quarter of the petroleum needs in New England and could be in operation by 1976.

New England has no oil refineries. The New Hampshire proposal is the first in that state and the third in the area. The others are in Maine. One is a plan for a $350 million refinery in Sanford by the Gibbs Oil Co. of Revere and the other is a proposal by Pittstown Oil for Eastport.

Booras said the Onassis company, Olympic Refineries, Inc., has international "contacts" that assure it a steady supply of Middle Eastern crude oil for the Durham refinery.

The oil would be unloaded at a marine terminal on the Isles of Shoals, east of Portsmouth, and then pumped through an underwater and underground pipe 15 miles to Durham Point.

Thomson, a long-time advocate of

an oil refinery for New Hampshire, said he was assigning an aide to spend the next six months clearing objections to the proposal. "We hope to convince the people of that area that a refinery will be an asset and will in no way be detrimental," said Booras.

Meanwhile, urgent preparations continued through New England for meeting the more immediate energy needs of this winter.

The president of the newly-formed Bay State Gasoline Dealers & Trades Assn., Harold J. Murphy Jr. of Springfield, urged dealers to close daily by 6 p.m. and limit sales to 10 gallons. He predicted that in addition to Sunday closings, these limitations will have taken effect statewide by mid-December.

"It's going to be real bad," said Murphy, explaining that the 15 percent cutback in deliveries ordered by President Nixon will take effect in December, traditionally the heaviest month for gasoline sales.

Automotive supply stores reported a booming business on locking gasoline caps. Managers at several stores said they were out of all popular makes.

FUEL, Page 12

"I've been in business 15 years and I've never seen anything like it," said Ralph Dow, assistant to the manager of Steven's Automotive in Wakefield. Two or three persons have already been in the store with stories of gasoline thefts from their vehicles, according to Dow.

Sen. George G. Mendonca (D-New Bedford) yesterday filed legislation requiring prison sentences for persons convicted of stealing gasoline from a motor vehicle. His bill calls for sentences of one to six months, which may not be suspended, and fines of $100 to $500.

"The gas shortage invites the stealing of this scarce commodity and at the same time poses the threat of severe hardship on the victims of such theft," Mendonca said.

Samuel Thompson, Boston Housing Authority administrator, breathed "a

large sigh of relief" after getting a promise of 400 barrels of heating oil for 13 projects which had been expected to run out today.

The oil, from C. H. Sprague Co. of Boston, is enough to last until Saturday, according to Thompson. He said Secretary of Consumer Affairs John R. Verani has given assurances of enough oil after that from several large suppliers to keep the 33,000 public housing tenants warm through December.

Thompson expressed hope that the mandatory allocation program enacted by Congress will be in operation by the end of the year to assure continuing supplies throughout the winter.

"*Onassis planning $600m refinery in New Hampshire.*" The Boston Globe, *November 28, 1973.*

Editorials

The overwhelming issue

1. The environment

If locating a refinery on the Great Bay is a bad decision in itself, nothing good can come of it if the Seacoast community or New Hampshire allows it to stand.

If the decision is a bad one from an environmental point of view then all of the assurances of the principals in this proposal will only amount to an effort to minimize a bad decision.

The way to handle a bad decision not yet in effect is to avoid it, rather than attempt to minimize its effects.

Therefore, the proper approach to this enormous proposal is not to seek assurances, but to stop the forward progress of the venture and take the necessary time to determine its value.

2. Planning

Regardless of its environmental effects from a conservationist's point of view, the refinery will have an enormous impact on local and regional planning.

For Durham, it would mean a complete uprooting of that town's master plan, because a refinery of this size would affect every aspect of town life. One can make an endless list beginning with housing, water and sewage facilities, roads, schools, businesses and so on.

To plan on ways of fitting the refinery into the seacoast community at the same time it is under construction is the same as no planning at all.

Therefore, the question of planning, like the ecological questions, demand that the forward progress of the venture be stopped so that necessary time be available.

3. Project immediacy

Obviously, as land was being cornered, the projectors of this refinery found it important to be canny, secretive and hasty.

Now that the project is out in the open, it is imperative that it be halted until the seacoast

community and Durham can study what is projected.

And if the project principals argue that they cannot wait because there are business or financial or some other imperatives, then the enterprise deserves to collapse.

The democratic process

This paper believes that the way in which this proposal was thrust upon the seacoast was outrageous.

Whether or not the proposal itself is outrageous will be defined

"IS THAT YOU, UNCLE MEL?"

as information is gathered and intelligence is brought to bear on the problem.

One solid impression gained from this week's massive public meeting of Save Our Shores (SOS) was that Durham seems as capable as any town in organizing the democratic processes that normally are capable of combatting a potential threat.

A second impression was the priceless resource the town has and the region has in the community of individuals that comprise the University of New Hampshire.

Two irrelevancies

The oil shortage

Whether Durham should want an oil refinery, or whether Durham should have an oil refinery, are questions that have nothing to do with the oil shortage.

Nevertheless, the refinery is being proposed in the context of the current declared crisis. And the crisis is being used to argue the refinery's virtue.

The subject of the nation's energy needs ought not to be

allowed to intrude into discussions of a refinery in Durham, because it is totally irrelevant. Obviously a person can be in favor of increasing our refinery capabilities by ten-fold and still disapprove of a refinery in Durham.

An oil refinery is a manufacturing plant. In this case, it would be one that draws on raw materials imported from other countries.

To suggest, as some are already doing, that it is the town's patriotic duty to give itself up to a refinery is absurd. The town is not sitting on a natural resource that needs to be developed. It is not a source of oil. It is simply one possible place where oil products might be manufactured. And this is not to say that it might be a good place to do it.

Although the subject of the oil shortage is irrelevant to the

Project Independence

question of a refinery on Durham Point, one aspect of this foolish juxtaposition of problems ought to be killed off immediately and forever.

A refinery on Durham Point, contrary to what Governor Thomson has said, would be harmful and not helpful to the President's proposal for Project Independence.

At the press conference announcing the refinery plans, Governor Thomson said:

"This can be good for New Hampshire, it can be good for New England and good for the entire Eastern Seaboard and this would be a concrete step in that second phase of the program that the President has announced as Project 1980 Independence."

What in fact is proposed by Governor Thomson is an oil refinery built by a foreign corporation or combine, drawing its oil from the very Arabian nations whose present embargo brought the oil shortage to our attention this past month.

In other words, a new manufacturing facility that would increase our dependence rather than free us of it.

Letters

Refineries and taxes

Editor:

I've seen several comments suggesting that construction of the proposed refinery on Durham Point could drastically lower Durham property taxes.

There has, however, been little discussion of the possibility that the proposed construction may not reduce Durham's taxes. One very possible action that would essentially eliminate Durham's prospective tax bonanza would be a decision by the State Legislature, by Congress, or by the courts that schools should not be supported by local property taxes.

If school support is shifted to a state property tax, the state of New Hampshire could enjoy a tax bonanza, but Durham wouldn't. In fact, the construction of an oil refinery could even stimulate Governor Thompson and other "friends of Durham" in the State Legislature to support a shift in school financing methods.

It would be ironic if residents of education-oriented Durham found

themselves encouraging their own representatives to fight against broader-based support for education in order to obtain a very low tax rate for themselves.

It seems to me that the possible tax advantage is very tenuous, and could easily be eliminated by the action of any one of several government agencies. The oil tanks themselves would not be tenuous at all.

Richard L. Kaufmann
Durham

Scoop praised

Editor:

Publick Occurrences may not be awarded a Pulitzer Prize for its scoop of the Refinery story but it surely deserves a public expression of gratitude. Its outstanding reporting on an issue that surely has become a cause celebre in the community, the state and the region. Congratulations!

John A. Beckett
Durham

PUBLICK
OCCURRENCES

Stephen A. Bennett — Publisher
Richard H. Levine — Editor
William F. Hansen — Advertising Manager
Win Suffern — Account Representative
Jay P. McManus — Reporter
Jay M. Smith — Reporter
Peter F. Fagley — General Manager
Robert C. Nilson — Artist

Published every Thursday by Durham Publishing, Inc. P.O. Box 398, Newmarket, N.H. 03857. Offices in the Gallant Mill Building, Newmarket. Phone 659-5538 and 659-5583. Single copy price 15 cents, yearly subscription $5.

Second Class Postage Application pending at Newmarket, N.H. Post Office

"Editorials." Publick Occurrences, December 7, 1973.

DECEMBER 13, 1973

Option at Shoals

Olympic Has

By JANE ANDREWS
Staff Writer

Olympic Refineries, Inc., has acquired an option to buy five-acre Lunging Island in the Isles of Shoals. The island area would serve as an unloading site for crude oil from supertankers for a proposed refinery on Durham Point.

The option was obtained this week from Mr. and Mrs. Robert Randall of Danvers, Mass.

In a telephone interview this morning, Mrs. Randall, a 50-

Island

year summer resident of the island and historian of the isles, said her lawyer told her it was "not impossible" for the land to be taken by eminent domain—that it had been done before elsewhere.

Olympic officials approached the Randalls four times when the couple, fearing the possibility that eminent domain might be used, decided to give them an option.

The selling price for the most westerly of the Shoals Islands was more than any offer they had ever received, Mrs. Randall said.

In an emotion choked voice, Mrs. Randall related that the family had struggled to keep the island which they have owned for the past 50 years.

They had battled with the Rye tax assessor for the last five years—especially after their taxes on the island went from $74 to $1,043 in one year, she said. Their taxes were finally abated to $843 a year. She said the family had written a number of letters to the tax assessors but hadn't even received a reply.

The Randalls had to borrow money for the assessment. "We spent every cent we had to keep it and now we are glad to lose it," she said.

She said that, in spite of the taxes, they didn't get protection for their property which was broken into three times last year.

When Peter Boeras, Olympic spokesman in New Hampshire, initially approached the Randalls, he told them he wanted to develop the island. "Strictly speaking, that isn't a lie," she said. The proposal struck her as strange since there is only one mooring site at Lunging and the cove is just about deep enough to float a shallow-draft skiff at "dead, low tide," she said.

Mrs. Randall said the family gave the matter a great deal of thought while they and their attorney checked into the issue thoroughly before they concluded they couldn't fight big business.

"We don't have any less love for that island or for the Shoals or for the work being done" in the marine laboratory on Appledore or for the conference center, she said.

Mrs. Randall worked as a waitress on the Iles when she was a teenager and has lectured about the history of the Shoals. She said she has been besieged by calls from people wanting to know what has happened.

"Olympic Has Island." Portsmouth Herald, *December 13, 1973.*

Numb. — 20

PUBLICK
OCCURRENCES

SAMPLE

Bulk Rate
Newmarket, N. H.

January 4, 1974   *The New Hampshire Seacoast Community*   **15 cent**

The art of Ikebana

Home heating in an emergency

Tommy Thompson's campaign

Why a farmer wants a refinery

Thompson's abandoned dairy farm

Can the State force an industry on a town?

"*Why a farmer wants a refinery.*" "*Can a state force an industry on a town?*" Publick Occurrences, *January 4, 1974.*

The Hampton Union   1/23/74

Page 12

Expression Of Opinion
Rye Residents Divided On Pipeline

By Lynette Finn

RYE - As legend has it, Blackbeard the Pirate left a beautiful golden haired maiden to guard his treasure and wait for him at Honeymoon Cottage on Lunging Island, one of the Isles of Shoals.

Some say they still see her, wandering and calling out, "He will come back." If and when Blackbeard does come back, perhaps he shall sail right by, not recognizing the place.

In a spot check survey taken to determine the concensus of opinion of town residents concerning the proposed Durham Point oil refinery which includes using the Isles of Shoals as an unloading station, the following statements were made:

"I am opposed to the proposed pipeline and refinery because industrialization threatens the preservation of our short seacoast and its wildlife. There are alternatives to this plan to provide fuel for northern New England. However, money and power may prevail and we should be knowledgeable about proper management of a pipeline if it must pass through our town."

"I don't see anything wrong with the idea. It will create jobs in the area. But it should be under American control. I think this whole shortage is a farce anyway because of big business."

"We should have an oil refinery in New England. There has not been a new one built since 1955. That is 18 years ago and think how the demand for oil has increased since then. The basic objection to one being here would be the oil spillage. I don't think the land area would be disturbed that much. Some people, I call them pseudo - environmentalists, have worked themselves up to a pitch. We should know more about the installation of such a project. We need facts, figures and a production scene."

"I am 100% against it. Especially in regards to what will follow the installation of the refinery. Chemical plants will be attracted. There goes our area completely."

"In my opinion, this will be just another pollution plant. I am definately against it. If I had any land, it would never be sold for this purpose. I'm surprised that Lunging Island has been put up for option. I do a lot of sailing around the Shoals and really would hate to see the area ruined."

"We are on the fence and need a lot more facts. All we know is what we read in the papers. The pipeline would probably still come through here even if Rochester gets the refinery. They owe us more information."

Another seacoast resident Donald J. Ryan, 68

Kings Highway. Hampton, expresses his feelings: "Should we ban air transportation, brain surgery, or the practice of anesthesia? No, because they have proven competant. Why should we assume that refinery technicians are less competent? Maybe we should ban educators, who only teach doubt of our own technology."

An organization called the Informational Group has been recently formed for the purpose of bringing the "facts and figures" to Rye residents.

"We prefer to keep the town on an open minded, unpolarized discussion basis rather than asking who is for and who is against the refinery," says Mrs. Louise Tallman of the Informational Group. "How can we evaluate the proposal without the proper information?"

"We are not an opposition group," she continues.. "It is important to work on all sides. We should be able to work with the Selectmen and deal with Rye on our own terms. Every town is in its own situation. We hope people will postpone opinions at this point until they get the whole picture."

It seems the townspeople want facts and figures. This meeting should be the time and the place. Who knows - maybe even Blackbeard will be there.

"Rye Residents Divided on Pipeline." The Hampton Union, *January 23, 1974.*

Oil spill would be disaster, 700 at hearing in Rye told

By R.S. Kindleberger
Globe Staff

RYE, N.H.—Scientists last night warned 700 residents of this coastal town that oil spilled from a proposed tanker terminal at the nearby Isles of Shoals could have a devastating effect on the marine environment.

A $600 million refinery for Durham was proposed in November by Olympic Refineries, Inc., which is owned by Aristotle Onassis.

The meeting in the Rye Junior High School gymnasium, was called to inform residents of the potential impact of the proposal on their community. Under the plan, crude oil would be unloaded five miles east of Rye at the Isles of Shoals, and then piped past Rye Beach to the refinery in Durham.

Plans call for the 400,000 barrel-a-day refinery to be completed in 1977.

"The question is not whether it will harm the environment, but how much" Dr. John M. Kingsbury, the director of the Shoals Marine Laboratory, said.

Kingsbury said studies have shown that crude oil and refined products even in minute quantities can be dangerous to lobsters and other marine life.

Prof. Frederick Hochgraf, a University of New Hampshire engineer, advised strengthening pipeline codes and warned of "special hazards" to underwater pipes.

In Durham, meanwhile, a town meeting set for March 5 will consider whether zoning bylaws should be changed to permit construction.

Hearings by the state's Bulk Site Review Committee are expected to begin next month. The committee is made up of representatives from 12 state agencies with responsibilities related to the refinery proposal.

Gov. Meldrim Thomson, a strong advocate of a refinery for New Hampshire, has promised quick action on the proposal.

Thomson also has said he will introduce a measure in the special February legislative session to assure that the New Hampshire Port Authority has the power to operate and regulate an offshore terminal.

A state report recommending a publicly operated terminal was endorsed earlier this month by Thomson. The report predicted that 10,000 refinery-related jobs would be generated 1985 and said millions of dollars in state revenue would result from a tax on crude oil flowing through an offshore terminal.

The possiblity of intensive development around a seacoast area refinery is one of the many objections raised by opponents.

Resistance is particularly stiff in Durham, where many residents fear the leisurely pace of the residential and college town would be permanently changed.

Critics in Rye, Hampton and other coastal tourist towns emphasize the pollution threat. "One spill could wipe out the entire tourist industry of Rye and Hampton," John Gibson of Save Our Shores said yesterday.

Hugh Gallen, Democratic candidate for governor, yesterday announced his opposition to a refinery anywhere along New Hampshire's 18-mile coastline.

"Our coastal resources are too few, too fragile and too important to the state's future to jeopardize," Gallen said. "Other sites are available and other communities are interested."

Robert Greene, consulting engineer from Dallas, Tex., said the possibility of pipeline leaks is "extremely remote." He said the danger of spillage on shore or in the water is "not something to be concerned about in any way."

Greene said his company chose the Durham Point site and still considers it an excellent site.

Several town officials questioned Greene and other consultants retained by Olympic and said many questions remain unanswered.

Greene said Olympic hopes to complete its studies and apply for permits in two to three weeks. Most questions from the audience were critical of the refinery proposal.

"Oil Spill would be a disaster, 700 at hearing in Rye told." The Boston Globe, *January 24, 1974.*

P.O. Box 364
Rye Beach, N.H.
Jan. 24, 1974

Malcolm Taylor, N.H.A.C.C.
5 South State St.
Concord, N.H.

Dear Tink,

      Harking back to the article on MASSACRE MARSH, you people may be interested to know that the proposed pipeline through Rye would pass directly through the Mill Field near the old Foss Mill Dam. In taking the picture of the Mill Dam which was used in the article, I was standing just about on the pipe strip.

      Talking with the lady who owns the tiny bit of saltmarsh which forms the Mill Field, and who did sign an option, her comments are of interest. "I was assured that there would be no damage to the ecology. They stated that the ground would be restored to its original state." I feel it depends on the point of view as to whether you consider there would be no ecological damage.

      A somewhat more modified statement was given by Mr. Greene of Purvin & Gertz, hired by Olympic Refineries. Rather than stating "no ecological damage", he stated that they have tried to choose a route which would minimize ecological damage. In asking advise of Fish & Game and others, this was directed toward adjusting the route. Very little detail would be attempted at this time until they know if they will be able to proceed in New Hampshire at all.

      I feel that there should be a description of the ecological losses of the land area of pipeline and refinery. Each Conservation Commission could go a long way on this in each town concerned. Our statements might come out a bit different from that hired by Olympic. Besides the Mill Field stated above, the line would pass through a fragment of climax forest, magnificent Beech trees of the same owner. These directly adjoin our Conservation acres. You can't put back 100-year old Beech Trees. If the Pipeline options lapse, it may be worth trying to figure how these acres of interest could be brought into preservation ownership.

      We had an outstanding Information meeting on the Refinery last night in Rye, covered widely by News media. Careful planning and a good moderator were the key. The concerned attention of a packed hall of 600 persons was impressive. Keppler of EPA so stated. Town boards, including Conservation, are studing the whole situation as a coordinated team. So complex!

      Sincerely yours,

Louise H. Tallman

Mrs. Louise Tallman's letter dated January 24, 1974, to Malcolm "Tink" Taylor, at the New Hampshire Association of Conservation Commissions.

Projection: 50% of current total

Refinery water needs could drink coast dry

The $600 million refinery planned by Olympic Refineries for Durham Point would demand about 50 per cent of the water now used by the entire seacoast region.

That much demand, says hydrologist Francis K. Hall of Durham, would speed the arrival of a drastic shortage of water for the area—and, by decreasing fresh water flow into Great Bay, endanger marine life.

Hall, a chairman of the Save Our Shores (SOS) Technical Committee, based his projections on statistics furnished last November by Purvin and Gertz, Inc., the Dallas consulting firm hired by Olympic.

The Purvin and Gertz report said that the refinery would need 6,000 gallons of water per minute for cooling—or 8,640,000 gallons per day (8.64 MGD).

The current average daily use of water by seacoast communities, Hall said, is about 16 MGD, with summer daily usage considerably higher and winter usage lower.

In discussing the possibility of a water shortage in the region, Hall pointed out that the largest drainage basin from which future water would come—the Piscataqua River Watershed—provides 14 MGD.

This watershed, covering an area of 700 square miles and including six rivers and their tributaries, serves all of Strafford and Rockingham counties with the exception of Seabrook and Hampton.

There are few possibilities, Hall said, of developing large reservoirs to store water which would extend the yield beyond 14 MGD.

The Oyster and Lamprey Rivers, which figure prominently in Olympic's plans, account for more than 40 per cent of the total water available from the Piscataqua Basin. The Oyster—with a peak summer flow of 1 MGD—and the Lamprey, with a peak summer flow of 12 MGD, couldn't provide enough water for the refinery during summer months.

In fact, it would take water provided by nearly one half of the total summer flow of the Piscataqua in New Hampshire," Hall said in the report.

Groundwater, desalination and reuse of sewage effluent are the only other possible sources of fresh water, Hall said.

"Groundwater provides 10.5 MGD. More groundwater might be developed," Hall said, "but in most cases, increased or new pumpage will deplete stream flows. Wells drilled into bedrock, he added, yield an average of only 0.01 MGD.

Although Hall calls desalination "technically feasible," he notes that in 1969 there were only three plants in the world with more than 7.5 capacity—one each in the USSR, Netherlands, and Mexico.

"A desalination plant to supply the proposed refinery with 8.64 MGD would be one of the largest in the world," Hall said.

But if the salt water percentage of the estuary is to be maintained, he said, the seawater input and wastewater output would have to be piped in from, and out to, the ocean.

"A major concern which applies equally to re-used or fresh water from local sources is that fresh water flow into Great Bay will be decreased, thus increasing salinity," Hall said.

This development, he warned, could have "serious biological consequences," and would also represent a total loss per day of more than eight million gallons of fresh water—leaving nothing for the future.

Although the 6,000 gallon-per-minute figure is the only one officially mentioned, Hall said refinery proponents have suggested informally that only 3,000 might be needed.

If so, this would simply reduce the shortage momentarily and introduce a delay before the shortage is really felt, Hall said. "But the overall consequences would be the same."

Rye harbormaster opposes oil terminal

John Widden, a lobsterman for 32 years and the harbor master for the town of Rye, opposes a supertanker terminal at the Isles of Shoals.

"Any oil spillage would end up right here on the beach because of the prevailing southerly and easterly winds," he said in a recent interview with *Publick Occurrences*.

"I don't think they can operate without spills," he said. "The best of operations are not foolproof."

Widden, who is paid $1,000 a year by the town to police Rye Harbor in his own boat, thinks that a supertanker terminal will cost the taxpayers money and hurt beach business.

"The town should have special clean-up equipment for oil-spills," he said. "We can't rely on the federal government or the Coast Guard to clean up an oil spill." He said that at present Rye has neither clean-up equipment nor the personnel trained to use it.

Widden also thinks that oil spills would lead eventually to property devaluation in all of Rye because the town would become a less desirable place to live or visit. Over half of the taxable residences in Rye are vacation homes.

But as a lobsterman, Widden is especially concerned about the threat to his fishing waters. He said that if blasting were necessary to dig a trench for the pipeline along the seabed, the concussion would "kill everything in the immediate area." If the pipeline were laid in a straight line from the Shoals to Rye, it would pass through the main lobsterbeds fished by the 60 or 70 full time and 300 or 400 part-time lobstermen in the area. The New Hampshire lobster industry produces about 135,000 pounds of lobster per year, Widden said.

"Refinery water needs could drink coast dry." Publick Occurrences, January 25, 1974.

Fishermen oppose oil on coast

Olympic Refineries Inc. brought its plans for a supertanker terminal at the Isles of Shoals to seacoast fishermen this week but the fishermen—after a two-hour presentation by the firm—appear to stand firmly opposed.

The N.H. Commercial Fishermen's Association, which represents most of those who take their livelihood from the sea, was scheduled to take a vote on the Isles of Shoals proposal last Tuesday night in Rye.

Without all the membership present, however, Chairman Robert McDonough postponed the vote until February 26. It doesn't seem to matter—association insiders say the majority vote will surely be 'No'.

With the exception of the Marconi brothers—who already decided to support the refinery plan—most fishermen at the meeting seemed dissatisfied with the answers they were getting.

"We are here to explain the benefits of the refinery to you as fishermen," said Peter Booras, Olympic frontman in New Hampshire. "The people coming here eat fish like everyone else—and I expect they will want to enjoy your products."

The retort from the back of the room was short: "If there are any around."

Booras is accustomed to chilly receptions. "There will be plenty around, I assure you. As a responsible applicant, Olympic intends to take steps to see that there are no ill effects to th environment."

At a brief bit of handshaking and grinning after the meeting, Booras was revived from a chat with the Marconi brothers—Mike, Joe and Geno—who assured him they were in favor of the project.

Before Booras arrived, in fact, it was Geno who chastized McDonough for passing on his personal views of the refinery to the press which—Geno said—naturally assumed McDonough was speaking for the fishermen.

"I live in Hampton myself a part of the year," Booras said, "and I certainly wouldn't want to see this area destroyed. But we shouldn't allow our fears to dominate decisions before all the facts are in."

When would the facts be in?

"In the coming weeks," Booras replied. Meanwhile, his presentation to the fishermen used phrases like "sincere objectives," "detailed studies," "integrity," "responsible applicant," and "in no way injurious."

Jeff Ritter of Rye

"Those are the facts, gentlemen," Booras said. "And we would like you to visit one of the most modern refineries in the world—in Toronto—to see for yourselves.

"The wastewater from that plant is clean. You could drink a cup of that water. In fact, the effluent waters from that refinery pass through an aquarium full of fish."

The fishermen listened politely. But neither Booras, nor Geno Harlow of Frederick R. Harris, Inc., were prepared to answer their questions about the specific impact of oil at the Isles of Shoals.

What about Olympic bonding to compensate fishermen for spills? "I really can't answer that. It's in the hands of the legal people."

And the necessary statutes and regulatory agencies to ensure the entire complex is properly built? "Several state statutes are being revised," Booras said. "But, in any event, Olympic would provide trained personnel to operate the Isles of Shoals facility."

How would the need for unloading equipment affect the coastline? "I can't forecast what the successful bidding contractor will do," Harlow said, "but I imagine there will be no great beachfront work in Rye."

What about the promise of jobs? Are there any contractors in southeastern New Hampshire qualified to build an oil refinery? "I can't say because I don't know," said Booras. "Olympic wouldn't consider laying pipe with inexperienced men. But I am sure, with a shipyard in the area, that there are many good welders and pipefitters."

If referenda in Rye and Durham say no to a refinery, will Olympic abide by that decision? "I would hope that the people would wait for the full facts before holding a referendum," Booras said.

Harlow, a professional engineer who helped design the port facilities in Portsmouth, was careful to answer only the questions he could answer.

The refinery will draw 600,000 Kw of power, Harlow told a questioner. And there is adequate power supply from the Public Service Company to supply that need—even if the Seabrook nuclear power plant is not built.

The refinery would need 3,000 gallons-per-minute of fresh water for cooling purposes at Durham Point, Harlow said. (Purvin & Gertz, the Dallas consulting firm hired by Olympic, said in a November, 1970 report, that 6,000 gallons-per-minute of fresh water would be needed.)

The discharge, with 10 ppm of pollutants, would be "so small as to have no effect whatsoever offshore," Harlow said.

What about the pipeline route from Isles of Shoals ashore? "A straight shot, with perhaps one bend," Harlow said. Time? "About 10 months." Adverse weather? "It shouldn't affect the construction time." Where will the pipe come ashore? "At Concord Point."

The fishermen were also skeptical about assurances from Olympic that the products from the refinery will be distributed to New Hampshire first—and the rest of New England second.

Eugene Harlow

"Why should Mr. Onassis sell his oil for 30 cents in New Hampshire when he can get 35 cents for it in New Jersey," asked one old timer.

"Well, the price structure works like this," Booras said. "The base is set at the refinery, and increases as it is distributed. New Hampshire is at the end of

Turn to Fish, page 13

"*Fishermen oppose oil on coast.*" Publick Occurrences, *January 25, 1974.*

Foster's Daily D

15c PER COPY

HOME DELIVERED 35c Per Week

Serving Southeastern New Hampshire and Southern

Geo. J. Foster & Co., Inc, Publishers

Second Class Postage Paid at Dover, N. H. 03820

DOVER, N.H., TUESDAY EVENING, FEBRUARY 12,

Slam Olympic Oil For Lack Of Homework

By BURT KEIMACH

NORTH HAMPTON — Speaking to a packed junior high school gymnasium, Kittery, Maine Planning Board member John Hallet took advantage of a rabidly anti-refinery audience as he blasted Aristotle Onassis and called the Greek shipping magnate's oil proposals a "disgrace."

"We're doing a lot of Onassis' homework for him. There's a lot wrong with that because he should have done it," Hallet said.

Continuing his blast, the Kittery banker proclaimed, "He [Onassis] doesn't know the dif-

ferences in temperature between Greece and New Hampshire." Hallet was alluding to what he called an original report of the Scottish plan that saw officials in Perrin and Gerts consulting firm of Dallas, Texas. That document in its first revelation to Onassis did not specify any particular area, for the location of a refinery and transportation operation.

Hallet said in that connection he would agree with, and completely support the recommendation of the New England governors who plan to allocate at least $700,000 for a study plan to determine where in New England a re-

finery would be best suited to all conditions.

This conforms in spirit to a Scottish plan that saw officials in that small nation of five million people undertake a complete survey of the entire coast. That action resulted from the initial establishment of regional planning policies, and was envisioned to help prevent devastation of not only the environment because of oil drilling, but to aid in the maintenance of Scottish civilization.

Hallet had sharp words for New Hampshire Governor Mel-

drim Thomson Jr., the only governor to buck the proposal for regional planning. "It's only common sense that we've got to work together on this. It appears, however, that Meldrim, (i.e.) does not want to sacrifice his options."

Olympic refinery officials were invited to the gathering, billed officially as an informational meeting of the Concerned Citizens of North Hampton, but sent word that they would not be able to attend.

That provided the opportunity for another Hallet blast, as the

Kittery man criticized the meeting planned for Durham on February 27, saying that voters would not really have enough time to digest the information presented only a few days before the March's town meeting.

"I don't wish Onassis any harm. But he's capitalizing on worry about oil. But since he was not done his homework his proposals are an insult to us all. We're not to that big a hurry. Let's take another look and see if we really need a refinery," he concluded.

There seemed to be a general

agreement that locating a refinery in this area, a place frowned upon by at least five major oil companies, would be the equivalent of building a city of 650,000 people with all its accompanying pollution and social problems.

Arthur E. Martin, also of Kittery, who served with Hallet as a member of the Committee for Regional Oil planning, outlined some of the problems with supertankers that contribute to their vulnerability.

He said that they are difficult

TURN To Page 12

MANCHESTER (N. H.) UNION LEADER — Thursday, February 7, 1974

Durham Resident Replies On Issue of Refinery

NELL E. CHAMBERLIN

Box 510, Durham

ESTABLISHED 1873

2, 1974

Olympic

to control because of the relatively small horsepower engines used to propel the massive ships. Martin also cited difficulties in stopping the tankers, or in turning them around without considering immense distances required for each operation.

Hitting another point that could cause a "Torrey Canyon" type of disaster, Martin said that the lee of the Isles of Shoals causes a "wave refraction." "This means that when large waves reach the shallow edges of the islands, they simply turn around toward its back, or lee side of the islands."

Charles Tucker, emphasizing that the group he directs, the Southeastern New Hampshire Regional Planning Commission, has not yet taken any official stand, predicted a huge "...economic boom for the area if a refinery is built. But then there will be a "bust when we become independent of Arabian oil, or it is cut off."

Tucker cited yesterday's nationalization of three oil companies by Lybia's Muammer Khaddafi. He also said that a refinery coming to the Seacoast might destroy the diversity that the area was finally beginning to have in industry and business, "...much as the naval shipyard in Portsmouth, simply because it was there, helped stifle the development of other types of activity here."

The meeting was replete also with an attack on Olympic Oil's Peter Booras. The Keene spokesman was chastised by University of New Hampshire chemistry professor G. Ulrick who said "The refinery will not be unobtrusive on the New Hampshire coast, as Mr Booras claims. It would be difficult to envision a shopping center and an apartment complex near that two square mile area.

Ulrich billed himself as a critic of Olympic. He outlined for the audience the actual chemical process of refining "crude oil" that might be needed, and expressed his concern over the question that might be needed, and expressed his concern over the question of whether or not the United States should become dependent or high sulphur Arabian crude oil.

"Just because we need oil, doesn't mean we need a refinery here. We need toothpicks and automobiles, but we don't necessarily make them here." Ulrich concluded.

"Slam Olympic Oil for Lack of Homework." Foster's Daily Democrat, *February 12, 1974.*

The Olympic presentation

p. 14–15

A critical omission

No mention of petro-chemical satellites

The Olympic refineries, Inc. presentation of its proposal for an oil refinery in Durham was marked by one serious, even critical omission.

The omission was of any mention of satellite industries and attendant petro-chemical industries.

Oil refineries, such as that proposed by Olympic, not only produce several products, but also create raw materials and waste which can be converted into useful products. There are many subsidiary industries which gravitate to refineries in order to shorten transportation distances and costs of raw materials.

To those people who know something about refineries, the matter of petro-chemical satellite industries are automatically associated with refineries themselves.

In fact, it is often this secondary, peripheral industrial growth which raises as much opposition in communities as the refineries themselves.

Furthermore, many communities have found that their legal abilities to control satellite industries are less effective than their abilities to control and regulate refineries.

There is a phrase for this growth of subsidiary industries. The phrase is "the landslide effect."

It was used by the U.S. Army Corps of Engineers in its interim report entitled "Atlantic Coast Deep Water Port Facilities Study" dated June, 1973.

That study concerned itself with port facilities to accommodate supertankers or the importation of oil and other bulk cargo on the Atlantic.

It concluded that measures of two types were needed "to ensure protection and enhancement of affected localities."

One was the reduction of the frequency and severity of oil spills by safety conditions attached to the construction and operating permits for the supertanker facilities.

Here is the passage from the Army Corps of Engineer study:

However, major landslide impacts could result from such facilities if they are not carefully controlled.

"Creating a point source for the importation of large quantities of crude oil could induce heavy concentrations of industrial facilities in areas having high environmental value, such as wetlands and recreational areas.

"Local interests have the ability to regulate the extent and nature of such growth through conditions applied to State permits and through local land use control.

"Nevertheless, historically, local governments have not demonstrated an ability to withstand pressures to use their lands for purposes of economic growth and development."

In discussing local attitudes towards a deep water supertanker facility, the report refers to a representative of the governor of New Jersey who "indicated a fear that potential landslide impacts could not be controlled and that the environmental and social costs associated with development of a terminal exceeded its potential benefits."

The report claimed that opposition on the East Coast to a deepwater port facility stemmed from fears of large oil spills at sea and "(2) induced industrialization of the hinterland—which, historically, local government has not been able to control."

This fear has not been mentioned by Governor Thomson.

However, last week, Congressman Louis Wyman, expressing his support for a refinery in New Hampshire, was particularly exuberant over the accompanying advent of petro-chemical industries which he looked upon, not with disfavor, but as a positive economic boon to the state.

One arm of New Hampshire government is aware of the significance of petro-chemical industrial growth: the office of the attorney general.

Edward A. Haffer of the Environmental Protection Division of the Office of attorney general, spoke earlier this month before a legislative fact-finding conference on oil refineries and offshore terminals.

Turn to CHEMICALS, page

Two ladies of Rye

thwart pipeline plans

Two elderly Rye women are blocking Olympic's effort to complete a pipeline right-of-way from the coast to Durham Point.

Frances Tucker and her sister Bernice Remick have refused repeated offers to buy their 55-acre farm on Brackett Road, which is directly in the path of the proposed pipeline.

Mrs. Tucker and Miss Remick refused the latest Olympic bid last week when Jeff Marple of Marple Associates, Portsmouth, working on Olympic's behalf, offered $200,000 for the property.

The offer was made by telephone, since the two sisters have long since stopped admitting real estate agents into their house.

After successive failures early last fall to persuade the women to sell, Olympic attempted to enlist the aid of Douglas R. Gray, chairman of the Rye Board of Selectmen, Reverend Richard L. Schlafer, minister of the Bethany Congregational Church in Rye, and Mrs. Tucker's daughter, who lives in Virginia.

Gray, who is a lawyer, acknowledged that he consented to talk to the women on behalf of a client, whom he understood to be representing a private individual. Gray also made a phone call to Mrs. Tucker's daughter, to ask her help, but she refused to cooperate.

Turn to PIPELINE page 2

The Olympic Presentation. Publick Occurrences, March 1, 1974.

the   committee has invited her and Douglas
...to meet with it next Wednesday in a
...closed session to discuss the matter
...further.

rporates... finally

...legal counsel, Hamblett, Kerrigan, La
Tourette and Lopez.

Gratsos is Onassis' head man in New
York and operates out of the Olympic
Airways office.

Papanicolaou works for another Onassis company, Victory Carriers, Inc.
and is a tanker chartering expert.

Lincoln is an in-house attorney for
Onassis' New York operations.

Hamblett is part of Olympic's New
Hampshire firm of attorneys.

Star Is. eyed
for radar station

Olympic's first choice for a tanker
radar communications station is on Star
Island but a member of the Star Island
Corporation says the island is not
available.

"We will not sell any of the island for
any price."

Barbara Rutledge of Dover is one of
the 200 members of the corporation
which sponsors religious conferences on
Star Island each summer. She has been

Booras phasing out

Peter Booras will become less important to the Olympic project as the need
for technical expertise increases, according to Nicolas Papanicolaou, company
vice-president.

Papanicolaou, in an interview with
Publick Occurrences after the Olympic
presentation in Durham, declined to
comment on the Attorney General's
investigation of Booras' campaign contributions in his unsuccessful bid for
U.S. Senator. Papanicolaou said that the
matter has nothing to do with Olympic.

Asked if Booras was losing favor in the
company organization as has been
rumored, Papanicolaou indicated only
that Booras will play a less and less
important role in the project as it moves
toward its technological phase.

Booras is the person who brought
Olympic and Governor Thomson together last year, and the person who
engineered the land acquisition for
Olympic which is still in progress.

Although Booras appeared briefly in
the crowd at the Olympic presentation,
he left early and was unavailable for
comment.

Attorney General Warren Rudman
said yesterday he would not reveal the
details of his office's investigation into
the 1972 Senatorial campaign contributions of Peter Booras.

He said such premature information
would be unfair to Mr. Booras and he
would therefore not comment on the
specific allegations under investigation.

attending such conferences for 26 years.

"People come from all over the U.S.A.
to attend the various conferences on the
island," she said.

According to Rutledge, the board of
directors has made a statement against
the Onassis proposal and its statement
reflects the opinion of the entire
corporation.

Asked if any member of the corporation supported the refinery, she replied

that there was not one member she knew
of who wanted to see the island used as
part of Onassis' plan.

Rutledge told the crowded audience
and the stage of Olympic consultants
that there was no way they would be
able to obtain land on Star Island.

She then told *Publick Occurrences*
that the only way Olympic would get
the island would be "over their dead
bodies."

"Star Island eyed for radar station." Publick Occurrences, *March 1,
1974.*

bordertown*NEWS*

Vol. 4, No. 10
Newburyport, Mass., March 8, 1974

PUBLISHED WEEKLY by the DAILY NEWS Newburyport

Olympic takes its case to the people

ISLES OF SHOALS

About six miles directly out to sea, this cluster of islands abounds in legend and history. Before 1614, when the famous Captain John Smith mapped the rocky and surf-lashed isles, early fishermen, traders and explorers had a part in their history.

THE ISLES OF SHOALS was partially hidden last week while the 84,000 dead weight ton tanker, the LaLabella, anchored between the islands and Rye. This is the approximate location of the two off-loading facilities planned by Olympic Oil Inc. for its 250,000 dead weight ton tankers.
(Buck Howe)

DURHAM, N.H. — The week that was for Durham began in Concord over the cries of power politics by local Durham residents.

Olympic Oil Inc., not yet incorporated in the state of New Hampshire, Monday, Feb. 25, unveiled in Concord a scale model of its proposed $600 million refinery and simultaneously fired off a 15-page booklet to the residents of Durham telling them, among other things, if the refinery is built, taxes in the town would drop from $47.80 per thousand valuation to $14.03 per thousand valuation or a drop of 70 per cent.

The critics of power politics came from a group of area residents and an official or two who felt Gov. Meldrim Thomson Jr. was using his office to promote a private concern, i.e., the refinery.

Despite the cries, the refinery model stayed in the State House two days before making its trip to Durham in time for the first of two informational public meetings.

It was during these meetings that the Durham public and other interested citizens were first briefed on what to expect from the proposed refinery. The meetings were held Feb. 27 and March 3.

The citizenry learned that Olympic Oil Inc. was totally owned by the Aristotle Onassis family and had been incorporated in New Hampshire Feb. 27.

The twin meetings held in the Lundholm Gym and Field House were televised over Channel 11, the university's national educational television station and was rebroadcast over the other NET stations in the state giving the entire state a look at the proposed refinery.

The meetings were guided by Alden Winn, chairman of the Durham selectmen. Chief spokesman for Olympic was Richard Greene, a vice president of Pervin and Gertz, the engineering firm hired by Olympic.

The company altered its presentation and question periods to give the audience a chance to ask questions on the presented material.

One point repeatedly brought up during the questioning was that of water issue. The company said

musical chairs

Let's talk about musical chairs, the political kind where we change philosophies not seats.

From a distance it seems we have something like that going on in Concord during the special legislative session and the tune they are all dancing to is called "Petro" (a Greek dance at that.)

Last week the dance centered around a bill introduced into the House of Representatives which calls for a local community to be it city or town to vote as to whether they would want a refinery to locate within their community.

"Home Rule" it was nicknamed and the dancing seems to be over who is indeed in support of that old yankee tradition of a community determining

Home rule or anti refinery?

By BUCK HOWE

CONCORD, N.H. — There was no doubt the refinery half of testimony because of a quorum call at the House Chambers. It resumed after a two-hour break with less

---

"*Olympic takes its case to the people.*" Bordertown News, *March 6, 1974.*

Can Rye stop a NH refinery?

The sign reads: OLYMPIC OIL OFF SHORE LOADING TERMINAL — NOW ZONED RESIDENTIAL — ISLES OF SHOALS

Andreae speaks for refinery, serves on refinery study group

Olympic says 'No comment'

Goode to edit UNH study

"Can Rye Stop a NH Refinery?" Publick Occurrences, *March 15, 1974.*

Editor:

The residents of Newmarket strongly supported the construction of an oil refinery in the seacoast area providing all State and Federal air pollution and environmental standards are met. The ballot vote was 575 YES and 355 NO.

The people of the state of New Hampshire also strongly supported home rule.

If the people of the entire state so overwhelmingly supported home rule, they should let the residents of Newmarket make their own decisions on the construction of an oil refinery in their own town.

The benefits gained economically would be tremendous to the town. The residents of Newmarket recognized this fact when they cast their ballots.

I hope that whatever company seeks to locate in Newmarket they will be given ample time to present their proposals to the residents. Let them tell us when they are ready to present their proposals.

Let the residents of Newmarket, not Durham, Rye, Exeter or any other community decide what is best for Newmarket. The people of Newmarket have indicated what their feelings are already.

Walter P. Schultz
Newmarket

Letter to the Editor in support of refinery in Newmarket. Publick Occurrences, *March 29, 1974.*

Olympic awaiting vote in Newmarket

Olympic Refineries is biding its time until the April 16 refinery vote scheduled for Newmarket.

According to Nicholas Papanicolaou, Vice President of the Onassis company, further decisions concerning the location of an oil refinery in New Hampshire or any other state will be based on prior local acceptance.

Olympic Refineries is still very much interested in New Hampshire, according to Papanicolaou. "There's a tremendous market and a tremendous need for a refinery there," he said.

When asked whether Durham was still a viable possibility, Papanicolaou said, "We're staying away from the whole Durham question right now."

Olympic Refineries will continue to consider any and all sites on the Eastern seacoast that express an interest in having a refinery, he said. He mentioned several southern states—Alabama, Mississippi and Texas—as possibilities. When questioned about the Lowell/Dracut, Massachusetts area, (recently mentioned as a possible target for a Venezuelan oil refinery), he said, "They've been in touch with us."

Olympic will also look into several other sites in New Hampshire besides Rochester, Newmarket and Durham, although he declined to say where because, right now, "they are such distant possibilities."

When asked whether Newmarket or Rochester would offer the better location for a refinery based on an engineering analysis, Papanicolaou replied that although the matter had not been studied in detail, he would be inclined to think that Newmarket would be preferable.

But, he added, further studies would be necessary for consideration of any new site.

"Olympic awaiting vote in Newmarket." Publick Occurrences, *March 29, 1974.*

Numb. — 32

PUBLICK
OCCURRENCE

March 29, 1974   *The New Hampshire Seacoast Community*   ...ents

Newmarket faces the refinery iss...

The new center

The Town of Newmarket now becomes the center of concern in the controversy over a Seacoast oil refinery.

Newmarket legally can reopen its voter checklist before the April 16 refinery referendum—a development that may aid anti-refinery forces (See below)

Olympic Refineries says it will mark time until the Newmarket referendum; also, Newmarket is preferred over Rochester. (Page 2)

Interviews with Newmarket's three selectmen and with leading local industries. (inside)

Newmarket Selectmen Hendricks, Vlodica and Schanda

Rolls may open for refinery vote

The Town of Newmarket may open its checklist to more voters before the April 16 referendum on an oil refinery within its borders.

Such an action might be significant, since it is believed by some people that anti-refinery votes would increase if the checklist were opened.

Newmarket voters, at town meeting, voted 575 to 359 in favor of a seacoast refinery and 523-401 in favor of one in Durham. Most people believe it to be a foregone conclusion that the voters would favor a refinery in their own town as well.

The governor's office has expressed an interest in the coming referendum. This week, an Olympic Refineries official said that his company would wait until the Newmarket vote before coming to any conclusions as to its future in New Hampshire.

Thus, a negative vote in Newmarket might end for good Olympic's plans in this state. A positive vote might keep Olympic's interest as well as stimulate the interest of other refinery companies.

Robert L. Stark, secretary of state, told *Publick Occurrences* this week that "the law is silent" on voter registration for special town meetings.

This would leave the decision up to the town's own supervisors of the checklist.

Stark's comments reverse the conclusion drawn from an earlier opinion expressed in a March 19 letter. At that time, he wrote "there is no provision for altering the checklist for special meetings other than for those where an election is to be held."

Richard Schanda, checklist chairman, took this to mean that the checklist could not be opened.

Stark explained to *Publick Occurrences* that he meant only that the town is not required by law to open the checklist; not that the town is prohibited from doing so if that is the desire of town officials.

Asked if he knew of any statutory prohibition to voter registration before a special town meeting, Stark acknowledged that he did not.

In his letter to Schanda, Stark referred to RSA 56:35. However, that statute deals with registration for primary elections only, not for special town meetings. Stark acknowledged this week that the statute does not appear to apply to this question.

Schanda, as well as the Newmarket selectmen, had interpreted Stark's letter to mean that no registration can occur before the important special town meeting to determine if Newmarket voters will accept a refinery within the town. They consequently dropped consideration of the idea.

Stark told *Publick Occurrences* he now recommends that the Newmarket supervisors of the checklist "obtain a legal opinion from their town counsel or from the courts," if they wish to pursue the possibility of holding registration.

Schanda, contacted for comment on Stark's new position, said that he will address another letter to Stark asking for a formal clarification of the matter.

William Beckett, town counsel, told *Publick Occurrences* that he cannot give an opinion on the appropriateness of voter registration for the special town meeting unless asked to do so by the selectmen.

No town officials were willing this week to speculate what effect, if any, the registration of new voters might have on the outcome of the refinery referendum, nor would they speculate on the number of new voters who would register.

However, members of anti-refinery groups, believe that reopening the checklist will bring in the "Stone Church" crowd, members of the younger class and the University of New Hampshire students living in Newmarket, who they think would be overwhelmingly opposed to a refinery.

According to the town clerk, Eileen Sadius, there are 1682 voters currently on the checklist.

Schanda noted that RSA 40:26 provides that people may vote at the special town meeting even though they are not on the checklist if (1) they voted in the 1972 biennial election, or if (2) they voted in the 1973 annual town meeting.

The town will vote on two questions on April 16.

1. Are you in favor of having an oil refinery located within the Town pro-

... applicable local, state and federal ordinances, statutes and laws, including but not limited to, all existing environmental, air and water pollution laws.

2. Are you in favor of requiring the Selectmen of the Town to call a special town meeting to approve any specific proposals made for the construction of an oil refinery at the Town of Newmarket.

Newmarket has already voted 575-359 in favor of a seacoast refinery and 523-401 in favor of a Durham refinery. Newmarket and Seabrook were the only towns in the Seacoast region to vote in favor of a refinery at regular town meetings in early March.

— Ron Lewis

Governor seeks welfare files

The Governor's Office has asked the State Welfare Division to turn over confidential personal files on about 150 welfare recipients attending institutions of higher education in the state, *Publick Occurrences* learned this week.

The Division of Welfare encourages qualified welfare recipients to enroll in institutions of higher education as a means of becoming self-sufficient.

Even though Thomas Hooker, the welfare director, is planning to turn over the requested information—and has already provided a list of names—the legal basis for the request is unclear.

Jay McDuffee, the Governor's press secretary, asked Deputy Welfare Director Thomas Thompson specifically for names and addresses of recipients, the amounts of individual welfare grants, child care allowances, allowances for books and fees, and amounts and sources of university tuition grants. The Division is now compiling the requested informa-

tion, according to a Division official.

Contacted on Wednesday, McDuffee would neither acknowledge nor deny that he had requested the personal files of welfare recipients, although, he did say that any request made by him to a government agency would be "made as an assistant to the Governor of New Hampshire."

McDuffee refused to answer any questions, saying that he would have no further comment until the source of the information was revealed.

Richard H. Craig, director of financial aid at the University of New Hampshire, refused to release tuition grant information requested by the Welfare Division last week.

"Information regarding any student's financial situation is considered confidential and can only be released with permission of the student involved," said in an interview.

Turn to WELFARE, page

"Newmarket faces the refinery issue." "The new center." Publick Occurrences, *March 29, 1974.*

Numb. — 37

PUBLICK
OCCURRENCE

May 3, 1974   *The New Hampshire Seacoast Community*

Music of ancient man	Spring gardening: organic & otherwise	UNH refin... tax chapter...
☒ ☒ ☒ ☒		

PSC's permit to pollute

•

The heifer and the coed

•

Seacoast gymnasts

•

Oyster River swim team

•

Art teachers exhibit

Olympic returns

Nicholas Papanicolaou

Taxing a refinery in Newington, Rochester and Newmarket

Local taxation of refineries is a big issue in Newington, Rochester and Newmarket.

John Davills, a resource economist and co-author with Douglas Morris of the tax chapter in the University of New Hampshire refinery report this week took a close look at the potential tax benefits of a refinery in each of these three communities.

In his view, Rochester and Newmarket could derive great tax benefits from refineries, but Newington would benefit very little, if at all.

In their chapter of the UNH refinery report, Davills and Morris did not discuss the question of refinery taxation in terms of particular communities.

Governor Thomson, who commissioned the UNH study, asked that the study be non-site specific. (See text of chapter, this issue.)

Newington has received a proposal from C. H. Sprague and Sons, Co. for a $10 million, 50,000 barrel per day refinery. Although the town turned down the idea of a seacoast refinery by a vote of 172-60 at town meeting, the Sprague proposal, if approved by the planning board, will be submitted to a vote at a special referendum.

Neither Rochester nor Newmarket has received a specific proposal, but both towns are attempting to attract refineries. Olympic Refineries, Inc., which failed to place a 400,000 barrel per day refinery in Durham, has announced its interest in both towns. In addition, Shaheen Natural Resources Company of New York has expressed interest in

Rochester

Rochester voted 1,386 to 986 in favor of a refinery at a special referendum on Wednesday.

Newmarket voted 644-526 in favor of a refinery on April 10.

In Newington, the tax rate "won't go down much, if at all," Davills said. He further predicted that if the refinery proposed by Sprague were to lead to a 25 per cent increase in town service costs by 1980, the tax rate could actually climb higher than without a refinery.

Based on Sprague's estimated building costs of $10 million, Davills figured that the refinery would add only $1.5 million in assessable property to the town.

If there were no increase in town services, the tax rate would drop about five per cent, he said.

Davills stressed that the amount of assessable property in a refinery complex is not clear under present statutes. His estimate of $1.5 million is only a rough guess, this figure is in line with the 15 per cent assessable valuation which Olympic Refineries estimated in its Durham proposal. But in Newington, where Sprague already owns most of the required storage tanks, the addition to the town's tax base could be less, perhaps under 10 per cent.

The refinery itself is considered a machine, and is thus exempt from taxation under state law. Also storage tanks connected to the main plant by pipelines are exempt. That leaves the administration building, workshops, land, separate storage tanks and perhaps

Turn to TAXATION page 2

New Hampshire is not Olympic's first choice

Olympic Refineries, Inc. is not jumping into any new refinery proposal until it is convinced that it has a decent chance of succeeding.

In a meeting last Tuesday, Nicholas Papanicolaou, Vice president of Olympic Refineries, Inc., Aristotle Onassis company, told the three Newmarket Selectmen that he would wait for the federal agencies to streamline their permit process and coordinate their policies before attempting to construct a refinery in New Hampshire.

Papanicolaou also asked for a show of support and a change of attitude in the Seacoast region.

Papanicolaou, in his opening statement to the town officials, said, "I asked to see you because as a company we would like to express our appreciation. But, we are not ready to make a proposal yet."

Papanicolaou said that he came to New Hampshire as a show of courtesy to the town of Newmarket.

Newmarket voted 644 to 526 in favor of an oil refinery on April 10.

"We are encouraged by the results but at the same time, we have to wait," said Papanicolaou.

The meeting was brief. The Olympic official politely thanked the Newmarket Selectmen for their town's show of support and answered questions concerning a refinery for Newmarket.

Papanicolaou said that if a proposal were to be made to Newmarket, probably it would be within 2 to 4 months if at all, and would remain essentially the same as the Durham plan.

The site would be 400,000 barrels a day and a deep water tanker terminal near the Isles of Shoals would still be a central part of the proposal.

But Papanicolaou continued to say "if", and there were no firm indications of any immediately forthcoming proposal.

In an interview after the meeting,

Turn to OLYMPIC page 3

"Olympic returns." Publick Occurrences, May 3, 1974.

FIGURE IV-4
OLYMPIC REFINERIES, INC.
PROPOSED PETROLEUM PRODUCT DISTRIBUTION SYSTEM
WITH DURHAM REFINERY

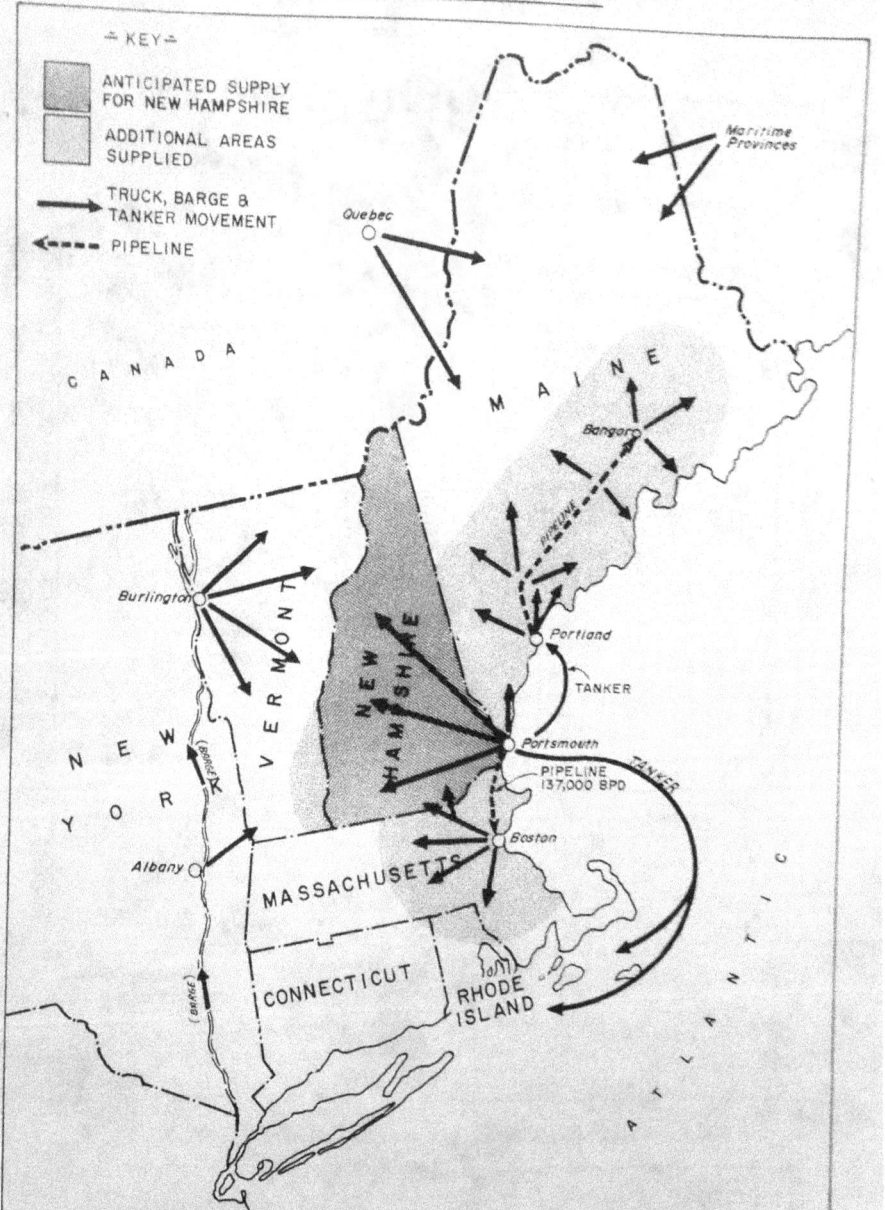

Proposed Petroleum Product Distribution by Olympic Refineries.

1691 Brackett Massacre Site. Photo by Lisa Moll, March 30, 2015

Conserved land along the proposed pipeline path. Rye, New Hampshire. Photo by Lisa Moll, March 30, 2015

Concord Point Beach Access. Rye, New Hampshire. Photo by Lisa Moll, March 30, 2015

CONCERNED CITIZENS OF RYE

STATEMENT OF POLICY

Rye is a recreational-residential community which offers clean beaches, boating and fishing to its residents and people who visit the seacoast area. We believe that an industrial oil refinery complex, with pipelines, offshore terminals and tanker traffic will destroy the irreplaceable natural resources of the seacoast and its offshore islands and will result in irreversible, heavy industrial change in the seacoast area.

For these reasons, the Concerned Citizens of Rye adopt the following policy statement:

(1) Public policies affecting Rye's future should be determined at the local level and in concert with surrounding communities.

(2) We recognize the need for sources of energy and urge that the people of New Hampshire support intelligent, coordinated New Englandwide planning to meet the energy needs of New Hampshire and the New England region.

3) We subscribe to the United States Environmental Protection Agencies' recommendations that heavy industrial oil refineries, pipelines and terminals should be located away from sensitive salt water estuaries and coastal areas.

(4) We oppose the stated plans of Olympic, Inc., for a refinery and offshore terminal system and will oppose any State of New Hampshire proposal which presents the same dangers to the quality of life in Rye and the seacoast area.

(5) We will take whatever action is appropriate in coordination with other area organizations of like mind to carry out this policy.

CCR's Statement of Policy

THE COAST WATCHER

Number One

February 12, 1974

Published by Concerned Citizens of Rye

Delegates in Exeter

Delegates from interested groups met Friday, February 8, for dinner at the Exeter Inn. Among these were Durham SOS Kittery's Committee for Regional Planning, the Seacoast Anti-Pollution League and Concerned Citizens of Rye. CCR was represented by Alan Pope and Judy Paterson.

These four organizations pledged support and agreed to meet periodically to compare and exchange facts and lend mutual assistance. Ms. Paterson explained the legislative Committee's lobbying work now under way in concert with surrounding towns and expressed our belief that Rye is presently the focus of contention.

Stirring to the South

An informational meeting took place Monday night in North Hampton, cosponsored by a citizen's group and North Hampton selectmen. It was similar to the January meeting at Rye Junior High School. Invitations were extended to Olympic's spokesmen. A panel of speakers included Arthur Martin and John Hallet of CROP, Dr. John Kingsbury of the Shoals lab and Charles Tucker of the South Eastern Regional Planning Commission.

Be Counted in March

The March 5 ballot will contain three informational questions and one proposed amendment to the zoning ordinance (the Tallman's amendment), which bear directly upon the oil issue.

Citizens may go register to vote in the Courtroom at Town Hall Feb.12,15,19 from 7-8 PM, and on Feb. 22 from 7-9 PM. Register and vote!

Active Support Needed

Those who wish to volunteer should see Betsy Bischoff at 31 Pioneer Road or call Betsy at 436-7892.

Contributions to the work may be mailed to Box 401, Rye, and checks made payable to Concerned Citizens of Rye.

GRAND OPENING!

The CCR Oil Facts Center will open this Sunday, February 17, at 2 PM, at 556 Washington Road opposite the Fire Station. Papers, posters, information and refreshments. Stop in. Tel.964-6514.

CCR Takes Stand

At a meeting last Sunday afternoon the staff of CCR voted unanimously to adopt and publish a statement of policy, the text of which reads:

Rye is a recreational-residential community which offers clean beaches, boating and fishing to its residents and people who visit the seacoast area. We believe that an industrial oil refinery complex, with pipelines, offshore terminals and tanker traffic will destroy the irreplaceable natural resources of the seacoast and its offshore islands and will result in irreversible, heavy industrial change in the seacoast area.

For these reasons, the Concerned Citizens of Rye adopt the following policy statement:

(1) Public policies affecting Rye's future should be determined at the local level and in concert with surrounding communities.

(2) We recognize the need for sources of energy and urge that the people of New Hampshire support intelligent, coordinated New Englandwide planning to meet the energy needs of New Hampshire and the New England region.

(3) We subscribe to the United States Environmental Protection Agencies' recommendations that heavy industrial oil refineries, pipelines and terminals should be located away from sensitive salt water estuaries and coastal areas.

(4) We oppose the stated plans of Olympic, Inc., for a refinery and offshore terminal system and will oppose any State of New Hampshire proposal which presents the same dangers to the quality of life in Rye and the seacoast area.

(5) We will take whatever action is appropriate in coordination with other area organizations of like mind to carry out this policy.

The Staff

Chairman	Peter Horne
V.Chairman	Sharon Fish
" "	Judy Paterson
Legal	Alan Pope
Outreach	Tom Barron
Information	Guy Chichester
Coast Watcher	Walter Wilson
Financial	Mike Flanagan
Technical	Kit Baker
Arrangements	Marion Barron
Treasurer	Bill Bischoff
Secretary	Kathy Baker
Legislative	Judy Paterson

The Coast Watcher. CCR Publication, February 12, 1974.

THE COAST WATCHER

Number Three —————— Published by Concerned Citizens of Rye ——————— March 2, 1974

The Bases

The Town of Rye will be two-hundred fifty years old in two years. If Olympic and its adherents have their will of us, in two years the town will be well on the way to becoming a heavy industrial site. Life in Rye will have been wrenched from an orderly development which has found room for fishing, farming, millwork, residence and recreation; it will be given over to support one gigantic petroleum plant.

In five years, if this comes, Rye will be a place where supertankers dock; a place where fleets of oil trucks pass on their distant missions of delivery; and in ten years, when satellite industries have arrived, as they inevitably must, there will be no distinguishing this town's good old face.

There is money at work in Concord, and reports of statewide majorities in favor of the thing. Those of us who oppose, and are now engaged in the legislative struggle, see a hard, uphill course ahead with only one thing certain: if we do not oppose, and that to the fullest extent, Olympic will shortly be pouring concrete in the woods of Durham Point and blasting fifty-foot pipeways in the ledges of this town.

CCR, though large and growing, cannot by itself defend Rye; the majority must determine such a course. We believe that the town's officials will heed the people's intentions, and if called upon, will stand at our head.

Questions A and C in Tuesday's ballot provide the legal and moral bases for a unified position.

The zoning amendment must pass or we lack any means to check development of Lunging Island or the waters off the Isles; we won't hear specific proposals or have the right to pass on the standards of their designs.

The three questions under section C of the ballot, though not legally binding, nonetheless will express the sense of the town; and sense is a prelude to action or the want of action.

The Olympic forces are moving fast, with Home Rule and Rye as major targets. We have a precious opportunity to speak on Tuesday. If you oppose radical transformation of the town, please vote "Yes" four times on Sections A and C.

The Wording

Questions pertaining to the oil issue appear on Tuesday's ballot in Sections A and C. These are reprinted below for the convenience of voters.

Question A is the zoning amendment, a short history of which appears on page two of this edition.

"Are you in favor of amending the zoning ordinance of the Town of Rye, NH, by the addition of that part of the Isles of Shoals in Rye as a single residence district?"

Question C is in three parts:

"1. Do you oppose off-shore loading and unloading facilities at the Isles of Shoals in connection with an oil refinery?

2. Do you oppose underground oil transmission lines transversing the Town of Rye?

3. Do you oppose an oil refinery in the seacoast area provided all environmental and safety standards are met?"

An article dealing with question C-3 appears on page two of this issue.

CCR recommends a "yes" vote in answer to each of the above.

Time to Vote

The polls will open Tuesday, March 5, at 10 AM, and remain open until 8 PM.

Babysitting will be furnished at Town Hall all day by the Rye Young Women's Club. Voters may leave children downstairs while they go up to vote.

COME ALL YE

CCR will host a report to the town on Monday, Election Eve, starting at 7:30 in the gym at the Junior High School.

The meeting will be composed of short speeches, a twenty minute slide shown on the Benecia refinery in California and how it really operates, and an open-ended question and answer period.

Guest speakers include Arthur Martin and John Hallet of CROP (Committee for Regional Oil Planning), Charles Tucker of the South Eastern Regional Planning Commission and John Horrigan, a UNH economist.

Refreshments will be served.

Question C-3

"Do you oppose an oil refinery in the seacoast area provided all environmental and safety standards are met?"

The phrasing of this ballot question implies that enviornmental and safety standards actually can be met. But CCR has concluded, in spite of promises by Olympic of a clinically clean operation with minimal destruction of nature that the reverse is true. There is no clean refinery presently operating in this country, and no likelihood of one in the near future.

The refinery in Benecia, California, cited by Olympic as a modern model, was built in 1968 by Exxon. One-fifth the size of Olympic's proposed plant and built at one-quarter the cost, the Benecia refinery discharges 24 tons of particulate matter into the air every day.

In December, 1971, Exxon was obliged to pay residents for damaged roofs, house paint and auto finishes caused by a control failure that covered Benecia with soot.

In May, 1972, the Vallejo Times reported "a holocaust that darkened the Northeast sky with a pall of oily black smoke." Exxon admitted it was the refinery's second fire in four years.

During one 18-month period, the refinery was issued 109 violation notices by local authorities, and during one 9-month period suffered 158 "upset" or breakdown conditions.

Oil is inherently an unclean business in its transportation, handling and especially its refining.

CCR has advocated regional oil planning and does not reject the idea of refineries out of hand; but this ballot question, framed as it is, is misleading and potentially harmful. Concerned Citizens of Rye, therefore, suggests that voters answer question C-3 with a "yes".

Thanks

There has been an immediate and gratifying response to our letter requesting donations.

Volunteers are still needed as well, and may join by calling Betsey Bischoff at 436-7892 or the Oil Facts Center, 964-6514.

Spills

According to the first report of the President's panel on oil spills, "The United States does not at this time have sufficient technical or operational capability to cope satisfactorily with a large scale oil spill in the marine environment. Nor does the capability exist to prevent virtually all of the oil in a massive spill from being deposited on shore". (Office of Science and Technology)

To Concord Again

Wednesday - Upwards of fifty Rye residents today joined two-hundred fifty seacoast and inland neighbors at a hearing on House Bill 18, the Home Rule Bill, before the Committee on Munici-

pal and County Government of the state legislature.

Lively speeches and a frank, democratic atmosphere crackling with applause, revealed the audience's strong support of the Home Rule principle; and an equal determination to oppose state propositions for a town which run counter to the will of that town's citizens.

North Hampton

Last Sunday some twenty-five residents of North Hampton, meeting at the historic Philbrick Tin Shop on Atlantic Avenue, formed an oil opposition group named North Hampton Concerned Citizens.

Patricia Aichele was elected chairman and Edward A. Dodd vice chairman.

Best wishes to our neighbors in their work.

CCR Board

John T. Dunfey has accepted a directorship on the board of directors of CCR.

Isles of Shoals Zoning
by Charles Tallman

In order to come within the jurisdiction of town zoning, a given area must be shown on the official zoning map of the town in which it lies.

When interest in local zoning issues heightened last fall, it was noticed that the Isles of Shoals in Rye, through an oversight, had been omitted from the present map, which dates from 1953. Further study showed inclusion of the Isles on the zoning map to be desirable.

A petition was, therefore, drawn which asks that the map be revised, with the Isles designated as an additional, single residence area. The petition was signed by fifty voters.

At the first Planning Board hearing to consider this petition, twenty persons spoke for it and none in opposition. It was brought out that the existing conference center and Federal Government installations are prior non-conforming uses.

The Robert Randall family who own Lunging Island and Mr. William Orcutt, representing the Star Island Corporation, attended the second hearing and endorsed the addition of the Isles to the map. Again no one spoke in opposition. The Planning Board is, therefore, recommending that the voters approve the change on the March 5 ballot.

The Coast Watcher. CCR Publication, March 2, 1974.

THE COAST WATCHER

----- Published by Concerned Citizens of Rye -----

Number four   March 16, 1974

Opinion

Allies

Town Meetings in Newmarket and Seabrook last week voted in favor of a seacoast refinery. Newmarket favored the Olympic proposal by 123, with some nine hundred voting. This represented about sixty per cent of those registered in the town.

Notwithstanding Durham's 9-1 rejection of the project, and strong votes in Rye, North Hampton, Hampton, New Castle, Exeter, Greenland, and by the Portsmouth City Council, the state's administration is casting about for a town or group of towns which will welcome the oil business.

A special Town Meeting has been called in Newmarket, to see if voters will approve a refinery there.

Olympic has momentarily fixed its ambitions on Rhode Island (perhaps feeling that the ground up this way has not proved so smooth as had appeared from helicopter view); but now letters have reportedly gone out from Concord to American oil firms, inviting them to look us over. Should Newmarket welcome a refinery, there can be small doubt that this administration would bend every effort to ensure a terminal and pipeline route to serve it.

Legal questions surrounding use of Rye's waters and lands are unresolved; but no one disputes the fact that Olympic holds options for most of a route through the town. It is also reported to have optioned the B & M right-of-way round the south end of Great Bay, thus affording potential linkage from Rye through Greenland to Newmarket.

It is in Rye's interest to support the Concerned Citizens of Newmarket, a newly-formed grass-roots group, like our own, though independent, which is working to achieve a majority in their special town meeting in April.

CCN is organizing under pressure; they have two and one-half weeks to swing the vote in a town where strong sentiments are felt pro and con.

Newmarket must certainly decide for itself; but as Rye remains focal in plans for oil in the seacoast, it will not be amiss to take keen interest in the outcome across the bay, and to pull for like-minded neighbors.

§

Please help pay for our lobbyist. 200 have given so far. CCR, Box 401, Rye.

Gazette

Saturday—The Rye Town Meeting voted unanimously to direct the Selectmen to send a letter to Olympic's representative in New York, stating Rye's intention to resist implementation of any plans for off-loading facilities in the town.

Sunday—Representatives from opposition groups in ten towns gathered here in the afternoon. Concerned Citizens of Newmarket was present; recent events in that town were discussed, together with consideration of upcoming Senate action on several oil-related bills.

Monday—CCR's Board of Directors met in the Library. There was discussion of the town's legal position.

Tuesday—Aides from the Governor's office came to Newmarket, and met with the Selectmen in executive session. A special town meeting to vote on a refinery in Newmarket was called for the first part of April.

Wednesday—Seacoast residents attended a Senate hearing on oil-related bills at Portsmouth High School. An administration official introduced an amendment to House Bill 34 aimed at strengthening the Port Authority's control of coastal waters. This drew ugly murmurs from the 300 present.

Thursday—Concerned Citizens of Newmarket held an open meeting in the town library. The group is seeking a date for voter registration before the special town meeting in April.

Inc.

The Concord legal firm of Upton, Sanders and Smith is leading steps to incorporate Concerned Citizens of Rye.

CCH

Concerned Citizens of Hampton has organized, composed of residents from beach and town alike. Edmund and Mary Loughlin will co-chair the group with Patricia Foss. Selectman James Fallon will head the Legal Committee, and John Dineen the Financial.

The Board

Directorships on CCR's Board have been accepted by Phillip Drake, Jesse Herlihy, F. L. Higginn, and Norman Kent. There are nine members presently on the board.

The Coast Watcher. CCR Publication, March 16, 1974.

UNH SCIENTISTS SAY . . .

Spills At Isles Of Shoals Facility Could End Up In Cape Ann Water

By Jane Andrews

DURHAM, N.H. — Aristotle Onassis' proposal to build a 400,000-bbl.-a-day oil refinery here touched off a state-wide controversy that sent petitioners and pollsters scurrying around New Hampshire collecting signatures as people took sides for or against the refinery.

Meanwhile, officials from the Greek shipping tycoon's Olympic Refineries quietly obtained an option on five-acre Lunging Island, the southwestern-most of the Isles of Shoals. Super-tankers weighing more than 250,000 tons would dock at the offshore unloading facility near Lunging and pump crude oil through underwater and underground pipelines to the Durham Point Refinery, according to Olympic's present plans.

On Dec. 20, Onassis himself flew over the Isles of Shoals and Durham Point in his jet, and later hovered over Durham Point in a helicopter, before he touched down in Bedford, N.H., for a brief press conference and reception and left.

Onassis and his consultants billed the refinery as pollution-free — "clean as a clinic" — and they said it would look "almost like a beauty parlor," and, when lighted up, "like a Christmas tree."

Advance billing aside, New Hampshire fishermen can't agree on what effect the refinery complex would have on fishing. Members of the N.H. Commercial Fishermen's Association planned to vote Jan. 22 on whether they want a refinery in the area.

As of mid-January, the association's president, Robert

P. McDonough, wasn't sure which way the vote would go. McDonough thinks that regardless of the outcome of the vote, "We'd have to go to Concord and let our views be known to the legislature." The state legislature meets in a special session starting Feb. 19, and several refinery-related bills are on the agenda.

A month ago, McDonough hadn't decided where he stood on the issue. He wanted more information, which he has since acquired from a series of public meetings, held throughout the N.H. seacoast region, by Olympic officials and their engineering consultants.

AGAINST PROPOSAL

"I don't speak as president of the association, but I'm definitely against the proposal personally," he said. "Incidental oil spills, which they can't possibly contain in rough weather, would have a bad effect on the fishing industry."

On the other hand, commercial fisherman Michael Marconi disagrees. "If I thought it (the refinery complex) interfered with my living, I'd be against it."

Marconi's mother Evelyn said she trusts the governor and the state to maintain safeguards against pollution, and she pointed out that if there is no fuel, "We can't go out and get fish," and that then there would be little reason to protect the fish.

McDonough raised the question of how much protection the Isles of Shoals would offer the supertankers — one of the reasons Olympic chose the Shoals as the terminal

site. The lobsterman's 25 years of experience have taught him not to expect wave protection from the Shoals when the wind blows over 40 knots. "There's no lee then," he said.

The Army Corps of Engineers ruled out the Isles of Shoals as a deep-water port when they completed a study of the Atlantic Coast from Eastport to Virginia last June, citing lack of local interest and environmental opposition as reasons for steering clear of the New Hampshire coast.

At the same time the Corps finished its study, two University of New Hampshire oceanographers concluded from their own year-long study that anything dropped, spilled or thrown overboard off New Hampshire's coast would drift south and shoreward well over 50% of the time, unless it sank to the bottom.

Assistant Professor Theodore Loder and graduate student Thomas Shevenell spent a year dropping plastic bottom drifters overboard to trace the direction of the bottom currents. Beachcombers and fishermen dragging the bottom recovered the drifters and mailed the attached plastic postcards back to UNH after filling in the date and location where they found the drifters.

The two scientists found the results of their study corresponded with similar studies of surface currents and they concluded:

"In the inevitable event of an oil spill anywhere near the Isles of Shoals, these coastal currents will probably carry the oil to the

(Continued on Page 24-A)

Durham Point

(Continued from Page 3-A)

coast from essentially Rye to Cape Ann. In addition, if any waste materials are pumped out to the Shoals areas, this material will possibly be carried toward the coast."

Of 567 bottom-drifters dropped, the researchers recovered 209, or 37%, of them, a higher percentage recovery, they said, than for other similar studies. They warned that the fate of the missing bottom-drifters must be taken into account when scientists interpret this information.

National Fisherman Publication. March 1974.

HB 18

STATE OF NEW HAMPSHIRE

In the year of Our Lord one thousand
nine hundred and seventy- four

AN ACT

requiring local approval prior to approval
of site plans for oil refineries.

Be it Enacted by the Senate and House of Represen-
tatives in General Court convened:

1 Local Approval Required.   Amend RSA 162-F by inserting after
section 1-a the following new sections:

162-F:1-b   Approval by Towns.   A site plan for an oil refinery shall
not be approved and an oil refinery shall not be located in any town which
does not have a zoning ordinance in effect without a vote of approval of
a majority of the voters present and voting on the question at an annual
or special town meeting called for such purpose.

162-F:1-c   Approval by Cities. A site plan for an oil refinery shall not
be approved and an oil refinery shall not be located in any city which does
not have a zoning ordinance in effect without a vote of approval as hereafter provided:

I.   A site plan for an oil refinery may be approved by a two-thirds
vote of the entire governing body of any city; or

II.   If the governing body of a city should vote to place the question
of whether or not to approve the location of an oil refinery in said city
on the ballot for referendum, they may place said question on the ballot to
be voted upon at any regular municipal or biennial election, or at a special

Home Rule. House Bill 18, 1974.

AMENDED VERSION

HOUSE BILL No.___18___

INTRODUCED BY: Rep. Dudley of Strafford Dist. 4

REFERRED TO: Committee on Municipal and County Government.

AN ACT   requiring local approval prior to approval of site plans for oil refineries.

ANALYSIS

This bill would require a vote of approval of a majority of the town's voters prior to the approval of a site plan for an oil refinery. It would require similar approval of two-thirds of a city council or a majority of the voters in a city prior to site plan approval of an oil refinery in a city. In a city, a referendum could be held if the city council votes to submit the question to the voters at a regular or special election, or if ten percent of the registered voters petition the city council to place said question on the ballot at a regular or special election. The city council would be bound by the outcome of the referendum.

Amended Version. "Home Rule." House Bill No. 18

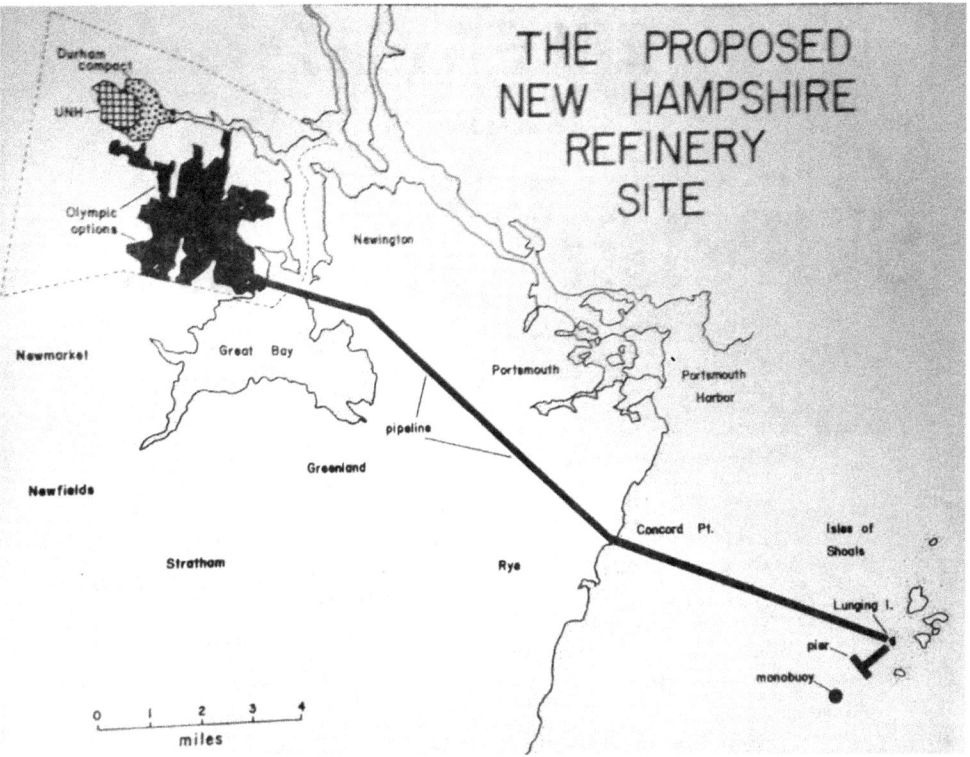

Diagram of the proposed New Hampshire refinery site. Olympic Refineries proposal, February 1974.

SAVE OUR SHORES TECHNICAL BULLETIN

A CITIZENS GROUP

CONCERNING THE SUPERTANKER FACILITY AT ISLES OF SHOALS*

There has been much concern for Durham Point and Great Bay arising from the proposal by Olympic Refineries, but there has been little attention paid thus far to the deepwater supertanker port at the Isles of Shoals. This part of the proposal poses a greater potential threat to the area than the refinery itself. The danger is twofold:

1) the threat of oil spillage and its accompanying contamination of the coasts of Maine, New Hampshire, and Massachusetts, and
2) the pressure for development of additional refineries and associated petrochemical and other "downstream" industries which depend upon refinery products.

While the oil spill danger is well known, we tend to think only of the comparatively rare catastrophic spills which are nationally reported. The smaller spills which constantly occur wherever oil is handled receive little publicity. Since 1970 the Coast Guard has been collecting data on oil spillage in U.S. waters. This data can be analyzed in order to assess just what kinds of oil spills can be expected from the proposed terminal. Since the amount of oil likely to be spilled is related to the total amount of petroleum products passing in and out of the terminal, any estimate of oil spillage must take into account the developmental pressures which serve to increase the demand and, therefore, the total throughput.

The pressures for development associated with a deepport are manifold. First is the economics of petroleum transport in supertankers themselves. Estimates of the savings involved in supertanker transport from the Persian Gulf, as opposed to transport in conventional tankers, run as high as $4.00 per long ton or 60 cents per barrel. It is clear that with potential savings of this magnitude the bulk of New England's petroleum imports will pass through this terminal. It is as a source of large quantities of economic raw materials that a deepwater terminal acts to spur industrial development.

In assessing the tendency for industrial development adjacent to a deepwater port, a study by Arthur D. Little, Inc. (ADL) for the Council of Environmental Quality states:

"The natural laws of economics would tend to increase crude oil throughout once the terminal facility has been constructed. As crude processing and directly associated industries increase the need for service industries would also increase as would the economic justification for locating them in proximity to the petroleum processes. This potential for major industrial and economic expansion associated with deepwater facilities is a prime reason why port development implications should be viewed from a long range (15-30) year time horizon."

The incentives for associated petrochemical development appears to be increasing. The same ADL study reports that the historical reliance of the petrochemical industries upon natural gas liquids is becoming economically unfeasible and it is expected that

*Researched and written by Professor Loren D. Meeker, Applied Mathematician, UNH

"Save Our Shores Technical Bulletin Concerning Supertanker Facility at Isles of Shoals." Undated.

p.2 (Concerning Supertanker Facility)

the industry will, in the future, draw heavily upon petroleum feedstocks. Speaking of this changing emphasis a Shell Oil publication, The National Energy Problem, Implications for the Petrochemical Industry, states "Economics,...,favor petrochemical complexes which are in close proximity to refineries which produce a full range of products."

Summarizing this aspect of the developmental pressures the ADL report concludes: "the construction of a deepwater terminal will therefore not only generate new refinery activity in proximity to the terminal facility, but will also tend to induce the establishment of large petrochemical complexes in the same vicinity, especially if the region already has a high degree of industrialization, if crude throughput volumes are large and if state and local land use controls are inadequate." (emphasis added)

Since New Hampshire, with a deepwater terminal, would satisfy two of the three criteria mentioned, the intense interest in the area by the oil industry becomes quite understandable. In our view, the oil industry has chosen this area to be the Linden, New Jersey of 1980.

Given these projections it is clear that the volume of petroleum products through a deepwater terminal at the Isles of Shoals will considerably exceed the 400,000-600,000 barrels per day (bpd) required for Olympic's refinery, wherever it may be situated. Certainly there would be a financial incentive for Gibbs Oil to reconsider their plan to transfer the crude oil for their proposed 250,000 bpd Sanford, Maine refinery from supertankers to conventional tankers at the Bahamas for delivery to Portland in favor of direct supertanker delivery to the Shoals. Such a decision would bring crude transport rates through the terminal to 600,000-800,000 bpd.

In addition to crude oil imports, a considerable portion of New England's demand for refined products (in excess of regional refining capacity) can also be expected to come inland via the deepport for trans-shipment by pipeline and surface transportation and for reloading on coastwide tankers and barges. All in all by 1980, the average daily throughput should easily exceed 1½ million bpd! It is quantities of this order of magnitude which must be considered when assessing the likelihood of oil spill damage.

The recent supertanker port studies (concerning sites at Machias Bay, Maine, and off the Gulf Coast of Texas) have assumed that in addition to chronic spillage of 30 gallons per operation one must expect a moderate spill of 500 tons (over 3500 barrels) annually and a catastrophic spill of 30,000 tons (roughly 210,000 barrels) approximately every 20 years.

Based upon the Coast Guard and worldwide oil spill data a group of MIT scientists have recently estimated the spillage rates arising from tanker operations. Under the assumption that spillage is proportional to total throughput this group estimates a spillage rate between 25 and 30 parts per million. That is, over one year's time they estimate total spillage from all causes to be 25-30 barrels spilled for each million barrels handled. These estimates are based on 1971 data. If we consider Boston Harbor we find, in a recent MassPort study by ADL and others, that the spill rate is in the range of 5 parts per million, or 5 barrels (nearly a ton) of oil spilled for each million barrels handled. In any case the message is clear when we consider the total volume to pass through the deepport. The terminal area will be subjected to chronic spill stress averaging in excess of a ton of oil per day!

There exist containment devices which in some circumstances will permit recovery of much of the oil spilled. The success of any recovery procedure is dependent upon the size of the spill, the speed of deployment of the containment barriers, and the sea and weather conditions. Authorities seem to agree that all operational containment devices begin to lose oil when currents exceed one knot and fail entirely in currents in excess

"Save Our Shores Technical Bulletin Concerning Supertanker Facility at Isles of Shoals." Undated.

p.3 (Concerning Supertanker Facility)

of two knots. Their ability to hold oil is also dependent upon the sea state. The best of them seem to be limited to sea heights less than eight feet. With these limitations it is certain that during a large portion of the year it will be impossible to contain much, if any, of the oil spilled at the terminal site. Since it is during such bad conditions that human and mechanical failures are most likely, we can conclude that large quantities of oil can be expected to escape from the terminal site and become subject to the currents and winds in the area.

At the completion of their recent current study in the Gulf of Maine, two UNH oceanographers described their results in local newspapers. They found that drift bottles released near the Isles of Shoals tend to reappear along the New Hampshire and Massachusetts coastline from Portsmouth to Cape Ann. We would anticipate much the same trajectories for oil spills emanating from the same region. The long term effects of this continual "oil bath" upon the region's tourist and recreational industries can be quite easily predicted; however, the effects upon the biological life of the shore are less certain. Evidence is now appearing that even one-time localized spills have a much longer impact on the creatures of the bottom than has been expected. For example, at Searsport, Maine, a small spill of jet fuel and #2 fuel oil (Coast Guard estimates were less than 75 gallons - under two barrels) occured in 1971. The Maine Department of Marine Resources found that more than half of the 23,000 bushel clam crop in Searsport's Long Cove were killed and that 8% of the clams which survived have cancerous tumors which did not appear to occur in clams outside the spill area. (Maine Times, Friday, December 21, 1973)

In summary we can state: A deepwater terminal at the Isles of Shoals will, because of the economic forces in the industry, act as a magnet for increased refinery capacity and associated petrochemical "downstream" industrial development. In addition to crude imports designated for local refineries, there will be considerable importation of refined products. Due to accidents, human error, and mechanical failures, there will be considerable oil spilled into the Gulf of Maine which, because of weather and sea conditions, will often be impossible to contain and recover. Much of this oil will impact upon the surrounding coastline. In the long run this chronic spill problem will have tragic effects upon the current recreational, tourist, and fisheries industries of the area.

The price is much too high for the citizens of New England to pay. With proper regional planning New England can have inland refineries serviced by low pressure pipelines and by medium to deepdraft tankers and still minimize the impact upon its priceless seacoast and estuarine systems. The refinery and its jobs and related new industries can exist without the destruction of its two most important coastwide industries, tourism and fisheries, and the suffering this needlessly entails for those making their living from the coast and the sea. All that is required is a cooperative regional approach to the problem and time to develop alternatives to this ill-conceived, spur-of-the-moment, opportunistic proposal put forward by Olympic Refineries.

"Save Our Shores Technical Bulletin Concerning Supertanker Facility at Isles of Shoals." Undated.

Concerned Citizens of Rye

Benefit dance

Peter Horne, Chairman

James Horrigan and Helen Reid

Betsy Bischoff

dy Jarvis

David Nixon, Horne (at top), Judy Patterson, cochairperson, Hugh Gallen, Harry Spanos, Nixon, Gallen and Spanos are gubernatorial candidates. More than 900 attended the dance at the Hampton Casino for the benefit of the group's legal fund.

Concerned Citizens of Rye Benefit Dance at Hampton Casino. Photo Credit: Judy Jarvis, Publick Occurrences.

Vol—2 Numb—3

PUBLICK
OCCURRENCE

August 30, 1974   The New Hampshire Seacoast Community   cent:

Seacoast groups fight Project Indepe   ice

The Boston Hearings

We have learned that wherever the industry goes it fouls the area's life-dependent resources.

—Guy Chichester, Citizens Coalition

...citizen remarks are being sought only as a formality and not with the intention of guaranteeing that citizen concerns will be implemented in project design.

—Nancy Sandberg of SOS

We feel that the six-year program of accelerated development to achieve self-sufficiency is a direct threat to our welfare, to our lifestyle, and to the immediate quality of our lives.

— Judith Paterson, Rye Concerned Citizens

Oil development in New England called damaging to the environment and dangerously hasty

Guy Chichester

"Seacoast groups fight Project Independence." Guy Chichester delivering remarks in the photo. Publick Occurrences, *August 30, 1974.*

Nashua Telegraph, Friday, September 27, 1974   3

Olympic Seeks to Construct Oil Refinery on Seacoast

By STEWART POWELL

CONCORD, N.H. (UPI) — Constantine Gratsos, president of Aristotle Onassis' Olympic Refineries, said today the firm hopes to announce a new site for a proposed oil refinery on New Hampshire's seacoast within one month.

"That's my estimate," Gratsos said in a telephone interview from New York.

Gratsos said the company filed plans for an off-shore crude oil terminal at the Isles of Shoals off the New Hampshire coastline two weeks ago with the Environmental Protection Agency (EPA).

He said the company was awaiting clearance from the EPA.

Olympic proposed last November to build the world's largest oil refinery at Durham, N.H.

The small college town rejected the project in March and the New Hampshire Legislature backed the town's decision with two bills on refineries and home rule during a special legislative session the same month.

Gratsos said Olympic is "absolutely" interested in making a refinery proposal in New Hampshire.

Gov. Meldrim Thomson has given strong support to the Onassis proposal and to all other suggestions for a refinery in New Hampshire.

Gratsos said his firm was working with the Army Corps of Engineers in an attempt to win approval for the offshore oil terminal.

He said the EPA had given "no commitment" about when the Olympic proposal would be completely reviewed.

Asked if Olympic still were interested in locating a facility in New Hampshire, Gratsos replied:

"Absolutely and positively. If we get clearance to do our project we'll do our project. If we don't get clearance, we won't."

Gratsos said he was in New Hampshire "about a month ago" to discuss the company's plans with officials whom he declined to name.

He emphasized the company would make the same proposal it made in Durham to one of three other seacoast area communities including Newmarket and Rochester. He did not name the third.

Like Durham, Newmarket and Rochester are some miles inland and would have to be fed by a major pipeline.

Both Newmarket and Rochester expressed support for a refinery in special votes following Durham's rejection of the proposal. Both are economically depressed.

"We are not making a new proposal," Gratsos said. "We are continuing our efforts of the past. That's all."

Gratsos' statement today was the first indication since spring that the company still is actively interested in locating in New Hampshire.

He said today, as he has in the past, that if New Hampshire were to reject the proposal, there are other communities in New England which have expressed interest in hosting a refinery.

Now You Know

More people are alive today than the total number of people who were alive and have died since the beginning of the Earth.

"Olympic Seeks to Construct Oil Refinery on Seacoast." Nashua Telegraph, *September 27, 1974.*

Notes

1. Loretta Britten and Paul Mathless, eds., *Our American Century Time of Transition: The 70s* (Richmond:Time-Life Books, 1998), 22-27. See also "Excerpts from Text of the President's Message to Congress on Energy," *New York Times*, January 24, 1974, 24. See Richard Nixon, "Address to the Nation About National Energy Policy," November 25, 1973, *The American Presidency Project*, http://www.presidency.ucsb.edu/ws/?pid=4051.

2. New Hampshire Department of Resources and Economic Development, George Gilman, Commissioner, *Economic Impact of Oil Refinery Location in New Hampshire*, December 1973. See also "Olympic Oil Proposal for New Hampshire Oil Refinery and Transshipment Terminal, Purvin & Gertz, Inc. Consulting Engineers, 1-2, November 1973, Oil Refinery papers, Oil Refinery Reports box, Rye Historical Society, Rye. For a good analysis of supertankers, see John M. Kingsbury, *Oil and Water:The New Hampshire Story* (Ithaca: Cornell University, 1975). See Loren D. Meeker, undated, "Concerning the Supertanker Facility at Isles of Shoals," *Save Our Shores Technical Bulletin*, 1-3, Oil Refinery papers, Oil Clippings 1973-1974, Miscellaneous Papers, box 1, Rye Historical Society, Rye. See Jay P. McManus, "Fisherman Oppose Oil on Coast," *Publick Occurrences*, January 25, 1974.

3. R.S. Kindleberger, "Oil Spill Would Be Disaster, 700 at Hearing in Rye Told," *Boston Globe*, January 24, 1974. Peter Horne, Rye town meeting moderator, telephone interview by author, April 7, 2015. John Whiteman, "Shoals Terminal

Stirs Rye Interest," *Portsmouth Herald*, January 24, 1974. John Kipner, "Onassis Refinery Plan Divides New Hampshire Area," *New York Times*, January 27, 1974.

4. "Rye, N.H., to get Pipeline?" *Lawrence Eagle-Tribune*, November 17, 1973. See also "Options on Durham Point for over 1,000 acres," *Publick Occurrences*, November 2, 1973.

5. Kingsbury, *Oil and Water*, 53. See also Dudley Dudley email to author, June 15, 2015, noting her recollection that the "telegram from William Simon to the governor was a forgery intended to persuade legislators and residents that the oil refinery would be beneficial."

6. Purvin and Gertz, Inc. Consulting Engineers, report to Olympic Oil, "Proposal for New Hampshire Oil Refinery and Transshipment Terminal," November 1973, Oil Refinery Reports box, Rye Historical Society, Rye.

7. "Text of Olympic Refineries-Onassis Press Conference in Bedford," *Publick Occurrences*, December 21, 1973. See also "Ari Inspects Site for New England's 1st Refinery," *The Daily News*, December 20, 1973.

8. "Onassis Refinery Plan Divides New Hampshire Area," *New York Times*, January 27, 1974. Nancy Sandberg, former chairman Save Our Shores, interview by author, Durham Historic Association Museum, Durham, March 4, 2015. Mel Low, Rye resident and opposition activist, telephone interview by author, March 30, 2015. Peter Horne, former chairman Concerned Citizens of Rye and Rye Town Meeting moderator, telephone interview by author, April 7, 2015. John Whiteman, "Shoals Terminal Stirs Rye Interest," *The Portsmouth Herald*, January 24, 1974.

9. "Expression of Opinion, Rye Residents Divided on Pipeline," *The Hampton Union*, January 23, 1974.

10. *The Coast Watcher*, February 12, 1974, Oil Refinery papers, 1974 Miscellaneous Papers, box 2, Rye Historical Society, Rye.

Peter Horne, chairman Concerned Citizens of Rye, telephone interview by author, April 7, 2015.

11. *The Coast Watcher*, Oil Refinery papers, 1974 Miscellaneous Papers, box 2, Rye Historical Society, Rye.

12. New Hampshire Department of Resources and Economic Development, George Gilman, Commissioner, *Economic Impact of Oil Refinery Location in New Hampshire*, December 1973.

13. Louise Tallman to Malcolm Taylor, N.H.A.C.C., January 24, 1974, letter, Oil 1974 Miscellaneous Papers, box 2, Rye Historical Society, Rye. See list of Rye opposition residents, edited July 15, 2015. See author's notes from meeting with Peter Horne, former chairman Concerned Citizens of Rye, and Alex Herlihy, vice president Rye Historical Society, Rye Town Museum, July 15, 2015.

14. "Two Ladies of Rye Thwart Pipeline Plans," *Publick Occurrences*, March 1, 1974. See also Thomas C. Clarie and Rosemary F. Clarie, *Just Rye Harbor, An Appreciation and History*, (Portsmouth:Peter E. Randall, Portsmouth Marine Society, 2005), 161.

15. "The Taking of Lunging Island," *Publick Occurrences*, February 1, 1974.

16. Kingsbury, *Oil and Water*, 31. See also "Star Won't Sell," *The Portsmouth Herald*, December 15, 1973.

17. Kingsbury, *Oil and Water*, 96.

18. "Questions and Answers Concerning Olympic Refineries, Inc.," Prepared by Purvin and Gertz, Inc. for Olympic Refineries, Inc. See also Kingsbury, *Oil and Water*, 45-47. Preliminary Study For Proposed Refinery," Volume 1 Summary, Prepared by Purvin and Gertz, Inc. for Olympic Refineries, Inc., Oil Refinery Studies 1,2,3 box, Rye Historical Society, Rye.

19. Dudley Dudley, former state legislator, interview by author,

Durham Public Library, Durham, March 13, 2015.

20. Kingsbury, Oil and Water, 45.

21. Kingsbury, Oil and Water, 46.

22. "Fisherman Oppose Oil on Coast," *Publick Occurrences*, January 25, 1974. See also "Preliminary Study For Proposed Refinery," Volume 1 Summary, Prepared by Purvin and Gertz, Inc. for Olympic Refineries, Inc., Oil Refinery Studies 1,2,3 Box, Rye Historical Society, Rye.

23. Kingsbury, *Oil and Water, pp.57-59.*

24. "Can Rye Stop a NH Refinery?" *Publick Occurrences*, March 15, 1974. The Town of Rye, State of New Hampshire, "1974 Reprint Town Warrant," Article 4, *Rye Town Reports 1973-1979*, Rye Public Library. See also *The Coast Watcher*, March 2, 1974, Oil Refinery Papers, 1974 Miscellaneous Papers, Box 2, Rye Historical Society, Rye.

25. University of New Hampshire, *The Impacts of an Oil Refinery Located in Southeastern New Hampshire: A Preliminary Study*, 1974, 29.

26. University of New Hampshire, *The Impacts of an Oil Refinery Located in Southeastern New Hampshire: A Preliminary Study*, 1974, III. Effects of Oil on Marine Organisms, pp. 50-52.

27. Kingsbury, *Oil and Water*, 51.

28. Kingsburgy, *Oil and Water*, 36. Nancy Sandberg, former chairman Save Our Shores, interview by author, Durham Historic Association Museum, Durham, March 4, 2015.

29. "N.H. House Upholds Home-Rule Zoning to Shatter Onassis Plan for Refinery," *The Boston Globe*, March 8, 1974.

30. Phyllis Bennett, former publisher *Publick Occurrences*, interview by author, Bennett residence, Durham, March 17, 2015. Dudley Dudley, former state legislator, interview by author, Durham Public Library, Durham, March 13, 2015. Nancy Sandberg, former chairman Save Our Shores,

interview by author, Durham Historic Association Museum, Durham, March 4, 2015.

31. "N.H. Legislature Tackles Problem of Regulating a Refinery," *Boston Globe*, February 24, 1974.

32. "Few Olympic Facts for Rye," *Publick Occurrences*, January 25, 1974. See also Kingsbury, *Oil and Water*, pp. 24-25.

33. University of New Hampshire, *The Impacts of an Oil Refinery Located in Southeastern New Hampshire: A Preliminary Study*, 1974. XIII. Institutional and Governmental Effects of An Oil Refining Industry in New Hampshire, pp. 38-39.

34. University of California, Santa Barbara, "1969 Oil Spill," http://www.geog.ucsb.edu/~jeff/sb_69oilspill/69oilspill_articles2.html.

35. U.S. Department of Energy,"The National Environmental Policy Act of 1969, as Amended," http://energy.gov/nepa/downloads/national-environmental-policy-act-1969.

36. U.S. Energy Information Administration, "New Hampshire State Energy Profile," accessed March 12, 2015 and March 9, 2016, http://www.eia.gov/state/print.cfm?sid=NH. See also "New Hampshire 10-Year State Energy Strategy," New Hampshire Office of Energy and Planning, September 2014 https://www.nh.gov/oep/energy/programs/documents/energy-strategy.pdf.

Bibliography

Carson, Rachel. *Silent Spring*. Boston: Houghton Mifflin; Cambridge: Riverside Press, 1962.

Kingsbury, John M. *Oil and Water: The New Hampshire Story*. Ithaca, NY: Shoals Marine Laboratory, 1975.

Schell, Jonathan. *The Time of Illusion*. New York: Vintage Books, Random House, 1976.

The University of New Hampshire, *The Impacts of an Oil Refinery Located in Southeastern New Hampshire: A Preliminary Study*, 1974.

Yergin, Daniel. *The Prize: The Epic Quest for Oil, Money and Power*. New York: Free Press, 2009.

Yergin, Daniel. *The Quest: Energy, Security and the Remaking of the Modern World*. New York: Penguin Press, 2011.

About the Author

Lisa Moll is a graduate student at the University of New Hampshire pursuing a Master of Arts degree in liberal studies. She lives in Rye, New Hampshire, with her husband and twin daughters.

www.ingramcontent.com/pod-product-compliance
Lightning Source LLC
Chambersburg PA
CBHW052213270326
41931CB00011B/2336